Terrestrial Environmental Sciences

Series editors

Olaf Kolditz
Hua Shao
Wenqing Wang
Uwe-Jens Görke
Sebastian Bauer

More information about this series at http://www.springer.com/series/13468

Agnes Sachse · Zhenliang Liao
Weiping Hu · Xiaohu Dai · Olaf Kolditz
Editors

Chinese Water Systems

Volume 2: Managing Water Resources
for Urban Catchments: Chaohu

Editors
Agnes Sachse
OpenGeoSys e.V.
Leipzig, Germany

Zhenliang Liao
College of Environmental Science
 and Engineering, UNEP-Tongji
 Institute of Environment
 for Sustainable Development
Tongji University
Shanghai, China

Weiping Hu
Chinese Academy of Sciences
Nanjing Institute of Geography
 and Limnology
Nanjing, China

Xiaohu Dai
College of Environmental Science
 and Engineering, UNEP-Tongji
 Institute of Environment
 for Sustainable Development
Tongji University
Shanghai, China

Olaf Kolditz
Department of Environmental Informatics
Helmholtz Centre for Environmental
 Research–UFZ
Leipzig, Germany

ISSN 2363-6181   ISSN 2363-619X (electronic)
Terrestrial Environmental Sciences
ISBN 978-3-319-97567-2   ISBN 978-3-319-97568-9 (eBook)
https://doi.org/10.1007/978-3-319-97568-9

Library of Congress Control Number: 2018933479

© Springer Nature Switzerland AG 2019
This work is subject to copyright. All rights are reserved by the Publisher, whether the whole or part of the material is concerned, specifically the rights of translation, reprinting, reuse of illustrations, recitation, broadcasting, reproduction on microfilms or in any other physical way, and transmission or information storage and retrieval, electronic adaptation, computer software, or by similar or dissimilar methodology now known or hereafter developed.
The use of general descriptive names, registered names, trademarks, service marks, etc. in this publication does not imply, even in the absence of a specific statement, that such names are exempt from the relevant protective laws and regulations and therefore free for general use.
The publisher, the authors and the editors are safe to assume that the advice and information in this book are believed to be true and accurate at the date of publication. Neither the publisher nor the authors or the editors give a warranty, express or implied, with respect to the material contained herein or for any errors or omissions that may have been made. The publisher remains neutral with regard to jurisdictional claims in published maps and institutional affiliations.

This Springer imprint is published by the registered company Springer Nature Switzerland AG
The registered company address is: Gewerbestrasse 11, 6330 Cham, Switzerland

Foreword

China's rapid economic development—coupled with strong population growth, increasing urbanization, and improved living standards—has led to an increasing burden on existing water resources in recent years. The supply and disposal structures were often unable to keep pace with growth, which among other things led to severe water pollution.

Germany has mastered similar water pollution challenges in its past. The use of modern environmental protection technologies and sustainable water management has resulted in globally recognized quality in water protection and drinking water quality. Experiences from two decades of intensive collaboration between research and industry can serve as an example for other nations to sustainably strengthen and improve environmental protection, especially in the area of vital aquatic ecosystems.

Sustainable water research has now become an important topic in the emerging economic nation of China. The Urban Water Resource Management (UWRM)-project presented here was able to analyze the current state of a Chinese urban and lake catchment with its practically applied methods and models and present treatment options. It not only raises public awareness of the issue but also brings together the representatives of international research institutes, local authorities, and global companies. The existing German-Chinese cooperation in research and industry should also be further expanded in the future in order to work effectively with current solutions to environmental problems.

Within this UWRM-project, the German nonprofit association OpenGeoSys e.V. is supporting the compilation of all project results in a compendium, and making it available to all interested parties. The purpose of the OGS e.V. is the promotion of science and research for computer-aided simulation in the environmental science and geotechnology. Those results often serve as a bridge to transfer knowledge between science, research, and the public. In addition, the association tries to promote other national nonprofit corporations in the field of science and research in ideal and financial ways. The purpose of the association is achieved by carrying out training courses, publicity work, advertising, and in particular by the production of visual material and publications like this book.

Leipzig, Germany   Thomas Kalbacher
March 2018

Acknowledgements

The project "Urban Catchments" is part of the German project cluster within the German-Chinese cooperation to the "Mega-Water Program" in China. It is funded by the Federal Ministry of Education and Research (BMBF) in the area of International Partnerships for Sustainable Technologies and Services for Climate Protection and the Environment (CLIENT) under the Research for Sustainable Development Framework Program. The funding under grant number 02WCL1337A-E is greatly acknowledged (Figs. 1 and 2).

Fig. 1 The "Urban Catchments"-project is funded by Federal Ministry of Education and Research of Germany

Fig. 2 Overview of all project partners

Contents

1 **Introduction** 1
   Agnes Sachse, Zhenliang Liao, Weiping Hu, Xiaohu Dai
   and Olaf Kolditz

2 **Managing Water Resources for Urban Catchments** 35
   Olaf Kolditz, Thomas U. Berendonk, Cui Chen, Lothar Fuchs,
   Matthias Haase, Dirk Jungmann, Thomas Kalbacher, Peter Krebs,
   Christian Moldaenke, Roland Müller, Frank Neubert, Karsten Rink,
   Karsten Rinke, Agnes Sachse and Marc Walther

3 **WP-A: Urban Water Resources Management** 87
   Peter Krebs, Firas Al Janabi, Björn Helm, Honghao Li,
   Benjamin Wagner, Christian Koch, Renyuan Wang and Lothar Fuchs

4 **WP-B: Development and Testing of a GIS-Based Planning Tool
   for Creating Decentralized Sanitation Scenarios** 125
   Thomas Aubron, Manfred van Afferden, Ganbaatar Khurelbaatar
   and Roland Müller

5 **WP-C: A Step Towards Secured Drinking Water: Development
   of an Early Warning System for Lakes** 159
   Marcus Rybicki, Christian Moldaenke, Karsten Rinke,
   Hanno Dahlhaus, Knut Klingbeil, Peter L. Holtermann, Weiping Hu,
   Hong-Zhu Wang, Haijun Wang, Miao Liu, Jinge Zhu, Zeng Ye,
   Zhaoliang Peng, Bertram Boehrer, Dirk Jungmann,
   Thomas U. Berendonk, Olaf Kolditz and Marieke A. Frassl

6 **WP-D: Environmental Information System** 207
   Frank Neubert, Matthias Haase, Karsten Rink and Olaf Kolditz

7 **WP-E: Groundwater Systems** 231
   Martin Pohl, Christian Engelmann and Marc Walther

Contributors

Firas Al Janabi Department of Hydrosciences, Institute for Urban and Industrial Water Management, Chair of Urban Water Management, Technische Universität Dresden, Dresden, Germany

Thomas Aubron Decentralized Wastewater Treatment and Reuse, Helmholtz Centre of Environmental Research–UFZ, Centre for Environmental Biotechnology, Leipzig, Germany

Thomas U. Berendonk Department of Hydrosciences, Institute of Hydrobiology, Chair of Limnology, Technische Universität Dresden, Dresden, Germany

Bertram Boehrer Department of Lake Research, Helmholtz Centre of Environmental Research–UFZ, Magdeburg, Germany

Cui Chen Department of Environmental Informatics, Helmholtz Centre of Environmental Research–UFZ, Leipzig, Germany

Hanno Dahlhaus bbe Moldaenke GmbH, Schwentinental, Germany

Xiaohu Dai UNEP-Tongji Institute of Environment for Sustainable Development, College of Environmental Science and Engineering, Tongji University, Shanghai, China

Christian Engelmann Department of Hydrosciences, Institute for Groundwater Management, Technische Universität Dresden, Dresden, Germany

Marieke A. Frassl Department of Lake Research, Helmholtz Centre of Environmental Research–UFZ, Magdeburg, Germany

Lothar Fuchs Institute for Technical and Scientific Hydrology, Hannover, Germany

Matthias Haase WISUTEC Umwelttechnik GmbH, Chemnitz, Germany

Björn Helm Department of Hydrosciences, Institute for Urban and Industrial Water Management, Chair of Urban Water Management, Technische Universität Dresden, Dresden, Germany

Peter L. Holtermann Department of Physical Oceanography and Instrumentation, Leibniz Institute for Baltic Sea Research, Rostock, Germany

Weiping Hu NIGLAS, Nanjing Institute of Geography & Limnology, Chinese Academy of Sciences, Nanjing, China

Dirk Jungmann Department of Hydrosciences, Institute of Hydrobiology, Chair of Limnology, Technische Universität Dresden, Dresden, Germany

Thomas Kalbacher OpenGeoSys e.V, Leipzig, Germany

Ganbaatar Khurelbaatar Decentralized Wastewater Treatment and Reuse, Helmholtz Centre of Environmental Research–UFZ, Centre for Environmental Biotechnology, Leipzig, Germany

Knut Klingbeil Department of Physical Oceanography and Instrumentation, Leibniz Institute for Baltic Sea Research, Rostock, Germany; Department of Mathematics, University of Hamburg, Hamburg, Germany

Christian Koch Department of Hydrosciences, Institute for Urban and Industrial Water Management, Chair of Urban Water Management, Technische Universität Dresden, Dresden, Germany

Olaf Kolditz Department of Environmental Informatics, Helmholtz Centre of Environmental Research–UFZ, Leipzig, Germany; Applied Environmental System Analysis, Technische Universität Dresden, Dresden, Germany

Peter Krebs Department of Hydrosciences, Institute of Urban and Industrial Water Management, Technische Universität Dresden, Dresden, Germany

Honghao Li Institute for Technical and Scientific Hydrology, Hannover, Germany

Zhenliang Liao UNEP-Tongji Institute of Environment for Sustainable Development, College of Environmental Science and Engineering, Tongji University, Shanghai, China

Miao Liu State Key Laboratory of Freshwater Ecology and Biotechnology, Institute of Hydrobiology, Chinese Academy of Sciences, Wuhan, China

Christian Moldaenke bbe Moldaenke GmbH, Schwentinental, Germany

Roland Müller Centre for Environmental Biotechnology, Decentralized Wastewater Treatment and Reuse, Helmholtz Centre of Environmental Research–UFZ, Leipzig, Germany

Frank Neubert AMC–Analytik & Messtechnik GmbH Chemnitz, Chemnitz, Germany

Contributors

Zhaoliang Peng NIGLAS, Nanjing Institute of Geography & Limnology, Chinese Academy of Sciences, Nanjing, China

Martin Pohl Department of Hydrosciences, Institute for Groundwater Management, Technische Universität Dresden, Dresden, Germany

Karsten Rink Department of Environmental Informatics, Helmholtz Centre of Environmental Research–UFZ, Leipzig, Germany

Karsten Rinke Department of Lake Research, Helmholtz Centre of Environmental Research–UFZ, Magdeburg, Germany

Marcus Rybicki Department of Hydrosciences, Institute of Hydrobiology, Chair of Limnology, Technische Universität Dresden, Dresden, Germany

Agnes Sachse OpenGeoSys e.V, Leipzig, Germany

Manfred van Afferden Decentralized Wastewater Treatment and Reuse, Helmholtz Centre of Environmental Research–UFZ, Centre for Environmental Biotechnology, Leipzig, Germany

Benjamin Wagner Department of Hydrosciences, Institute for Urban and Industrial Water Management, Chair of Urban Water Management, Technische Universität Dresden, Dresden, Germany

Marc Walther Department of Environmental Informatics, Helmholtz Centre of Environmental Research–UFZ, Leipzig, Germany; Department of Hydrosciences, Institute for Groundwater Management, Professorship of Contaminant Hydrology, Technische Universität Dresden, Dresden, Germany

Haijun Wang State Key Laboratory of Freshwater Ecology and Biotechnology, Institute of Hydrobiology, Chinese Academy of Sciences, Wuhan, China

Hong-Zhu Wang State Key Laboratory of Freshwater Ecology and Biotechnology, Institute of Hydrobiology, Chinese Academy of Sciences, Wuhan, China

Renyuan Wang Institute for Technical and Scientific Hydrology, Hannover, Germany

Zeng Ye NIGLAS, Nanjing Institute of Geography & Limnology, Chinese Academy of Sciences, Nanjing, China

Jinge Zhu NIGLAS, Nanjing Institute of Geography & Limnology, Chinese Academy of Sciences, Nanjing, China

Acronyms

ACLERP	Anhui Chao Lake Environmental Rehabilitation Project
ADaM	Aachener Daphnien Medium
ADB	Asian Development Bank
AEPB	Anhui Environmental Protection Bureau
AMC	AMC-Analytik & Messtechnik GmbH Chemnitz
AWATOS	Autarkic Water Observation System
bbe	bbe Moldaenke GmbH
BMBF	Bundesministerium für Bildung und Forschung, Deutschland, German Federal Ministry of Education and Research
cap	Capita
CAS-HYB	Chinese Academy of Sciences, Institute of Hydrobiology
CAWR	Centre for Advanced Water Research
CD	Capacity Development
Cd	Cadmium
CLIENT	International Partnerships for Sustainable Technologies and Services for Climate Protection and the Environment
CLMA	Chao Lake Management Authority
Co	Cobalt
Cr	Chromium
Cu	Copper
CRAES	Chinese Research Academy for Environment and Science
DDT	Dichlorodiphenyltrichloroethane
DDX	DDT-related contaminants
DEM	Digital elevation model
DFG	Deutsche Forschungsgemeinschaft, German Research Foundation
DiMoN	Application tool for statistical disaggregation of precipitation over time
DLR	Deutsches Zentrum für Luft- und Raumfahrt
DMP	Data Management Portal
DSM	Digital surface model

DWA	German Association for Water, Wastewater and Waste
DWSA	Drinking water source area
EC50	Half maximal effective concentration
ECMWF	European Centre for Medium-Range Weather Forecasts
EIS	Environmental Information System
ENVINF	Environmental Informatics
ESA CCI	European Space Agency Climate Change Initiative
EU-WFD	EU Water Framework Directive
EXTRAN	Explicit hydrodynamic transport model
FOG	Areas and objects in GIS
FTP	File Transfer Protocol
FYP	Five Year Plan
GDP	Gross domestic product
GETM	General Estuarine Transport Model
GIS	Geographical Information System
GMSH	Finite element mesh generator
GUI	Graphical user interface
Hg	Mercury
HYSTEM	Hydrological runoff model for urban drainage
HOCs	Hydrophobic organic contaminants
IARC	International Agency for Research on Cancer
IDW	Inverse distance weighting
IHB-CAS	Institute of Hydrobiology, Chinese Academy of Sciences
IOW	Leibniz Institute for Baltic Sea Research
itwh	Institute for Technical and Scientific Hydrology, Hannover, Germany
KDE	Kernel density estimates
LANDSAT	Multispectral data of the Earth's surface on a global basis
MET	Metoprolol
MIP	Multum in parvo, meaning "much in little"
Mn	Manganese
MoST	Ministry of Science and Technology of PR China
NaN	Not a number
NetCDF	Network Common Data Form
Ni	Nickel
NIGLAS	Nanjing Institute of Geography & Limnology, Chinese Academy of Sciences
NSE	Nash-Sutcliffe Efficiency
NSFC	National Natural Science Foundation of China
OGS	OpenGeoSys
OGS DE	OpenGeoSys Data Explorer
PAH	Polycyclic aromatic hydrocarbon
PCP	Pentachlorphenol
POP	Persistent organic pollutant
PRC	People's Republic of China

QA/QC	Quality assurance and quality control
QGIS	Quantum GIS
R&D project	Research and Development project
RECEIS	Research Centre for Environmental Information Science
RMSE	Root Mean Square Error
RSR	RMSE-observation standard deviation ratio
SDI-12	Serial Digital Interface at 1200 baud
SIGN	Sino-German Water Supply Network
SINOWATER	Good Water Governance-Project
SME	Small and medium-sized enterprise
SRTM	Shuttle Radar Topography Mission
SWMM	Storm Water Management Model
TAN	Total ammonia nitrogen
TBBPA	Tetrabromobisphenol
THMC	Thermo-Hydro-Mechanical-Chemical processes
THMs	Trihalmethanes
TI	Toxic index
Ti	Titanium
TIN	Triangulated irregular network
TU	Technische Universität
TUD	Technische Universität Dresden
TUD-HYB	Technische Universität Dresden, Department of Hydrosciences, Institute of Hydrobiology
TUD-SWW	Technische Universität Dresden, Department of Hydrosciences, Institute for Urban and Industrial Water Management
TVD	Total variation diminishing
UC	"Urban Catchments"-project
UFZ	Helmholtz-Zentrum für Umweltforschung
UV	Ultraviolet
UV-A	Ultraviolet radiation with a range of 315–380 nanometres
UV-LED	Ultraviolet light emitting diode
UWRM	Urban Water Resources Management
V	Vanadium
VGE	Virtual Geographic Environment
VR	Virtual reality
VTK	Visualization Toolkit Format
WISUTEC	WISUTEC Umwelttechnik GmbH
WKDV	Wissenschaftliche und Kaufmännische Datenverarbeitung, UFZ
WRF	Weather Research & Forecast Model
WWI	Waste Water Infrastructure
ZEBEV	Time coefficient method

Chapter 1
Introduction

Agnes Sachse, Zhenliang Liao, Weiping Hu, Xiaohu Dai and Olaf Kolditz

1.1 Idea of the Book

This book synthesizes the results of research into water research management conducted in university-based or affiliated institutes, companies and local authorities and their partner. The background is a research and development programme on "Managing Water Resources for Urban Catchments". It ran between 2015–2018 and was funded by the German Federal Ministry of Education and Research (BMBF). Scientists and company employees worked together in the inter-disciplines and international project. The overall goal of the project was the development of sanitary and

---

A. Sachse (✉)
OpenGeoSys e.V, Lampestraße 5, 04107 Leipzig, Germany
e-mail: agnes.sachse@web.de

Z. Liao · X. Dai
College of Environmental Science and Engineering,
UNEP-Tongji Institute of Environment for Sustainable Development, Tongji University,
Siping Road 1239, Shanghai 200092, China
e-mail: zl_liao@tongji.edu.cn

X. Dai
e-mail: daixiaohu@tongji.edu.cn

W. Hu
NIGLAS, Nanjing Institute of Geography & Limnology,
Chinese Academy of Sciences, 73 East Beijing Road, Nanjing 210008, China
e-mail: wphu@niglas.ac.cn

O. Kolditz
Department of Environmental Informatics, Helmholtz Centre
of Environmental Research–UFZ, Permoserstr. 15, 04318 Leipzig, Germany
e-mail: olaf.kolditz@ufz.de

O. Kolditz
Applied Environmental System Analysis,
Technische Universität Dresden, Dresden, Germany

© Springer Nature Switzerland AG 2019
A. Sachse et al. (eds.), *Chinese Water Systems*, Terrestrial Environmental Sciences,
https://doi.org/10.1007/978-3-319-97568-9_1

environmental engineering system solutions for sustained water quality improvement in the city of Chaohu. It involved an innovative approach—Urban Water Resources Management (UWRM)—that provides for efficient sanitary systems in urban and rural areas as well as with the needs of natural aquatic ecosystems. Chao Lake plays a central role as the ecological and economic resource and drinking water source for Chaohu City to be protected for future generations. This Research and Development Project (R&D Project) contributes greatly to the sustainable development of the Chaohu region and Anhui Provincial Government's master plan "Ecological Lake City of Chaohu".

1.2 Process and Character

The book is based on the results of five Work Packages (WP-A, WP-B, WP-C, WP-D and WP-E) dealing with Urban Water Resources Management, Decentralised Wastewater Management, Early Warning Systems for Surface Water Systems, Environmental Information Systems and Groundwater Modelling Systems. The book has a special character: It presents practise- and implementation-oriented results from the Sino-German cooperation and it shows how such cooperation can work and also reveals the challenges in international cooperation.

1.3 Target Groups

The book is intended to be used for three main target groups:

- Urban Water Resources Management Practitioners, for examples regional water authorities or water management institutions
- Scientists and funders of water resources management projects and those who are facing the challenge of implementation-oriented and interdisciplinary research
- Project members of the "Urban Catchment", to have the full project results summarised in one compendium

We hope that experience from Managing Water Resources for Urban Catchment project can be useful to them. And we hope that we can encourage others to proceed the previous research approaches and initial implementations and solutions strategies.

1.4 Overview

This book is the compendium (complete publication) of all working packages (WP) within the Urban Catchment project. It focuses on reproducibility as well as transferability to other regions (concepts, methods, workflows, application creation, etc.). The book supports the publicity work of Sino-German Research project as well as national and international project visibility. It includes various media from the three-year project period: presentations, abstracts, conference papers, photos and maps. The book presented here was compiled in close cooperation with the OpenGeoSys e.V.

This chapter describes the idea behind the book and introduces the Chinese research area around Chao Lake and the city of Chaohu, one of the fastest growing urban areas in the world. It also includes a detailed literature research about current issues in environmental research and describes the aspects of the 13th Five-Year-Plan of Major Water Project in Chaohu Basin.

In Chap. 2, the objective as well as the proposed solution and background of the German-Sino research project in Urban Water Resources Management are explained in detail. In addition, all German and Chinese project partners from universities, authorities and companies are briefly introduced. The chapter is rounded off by an overview of the temporal development of the 3-year research project. The methodology to achieve the project goal deploys the Urban Water Resource Management (UWRM) concept – holistically viewing the urban water network and all levels of the aquatic system according to the principle of emissions (source of pollution)/immissions (contamination).

Chapter 3 presents the results of Work Package A (WP-A), which deals with urban water resources management in the city of Chaohu and its surroundings. The city and the surrounding area draw their drinking water from the heavily polluted Chao Lake. In order to propose effective methods for improving water quality, a comprehensive analysis of the sea-city system is necessary, as well as the establishment of an online monitoring system for water quality (urban monitoring). Among other things, the chapter focuses on the available data sets (data scarcity especially regarding input data: precipitation) for hydrodynamic models and describes methods of data processing (e.g. disaggregation of rain data by using DiMoN tool). This pre-processing of input data is crucial for an effective drainage system assessment with hydrodynamic models (e.g. itwh.Hystem-Extran) and offers the possibility to present solutions in urban water management. An another part of the chapter presented the results of the generation of a digital elevation model (DEM), which was necessary for the delineation process of hydrological catchment areas. The Institute for Urban and Industrial Water Management (TU Dresden) and the Institute of Urban and Industrial Water Management (itwh Hannover) participated in this working package.

The catchment area of the Chao Lake is characterized by the diffuse sources of pollution by indirect discharges of untreated or inadequately treated domestic wastewater, pollution from livestock production and agricultural runoff and groundwater passage. Scientists from the Department of Decentralized Wastewater Treatment and Reuse (UFZ Leipzig) developed within the WP-B (Chap. 4) a comprehensive Geographical Information System (GIS) tool called "ALLOWS" to create and evaluate decentralised wastewater management scenarios that will be compared on a cost-efficiency basis in order to point stakeholders towards appropriate wastewater management solutions. Using GIS enables the visualisation of the information, highlighting particular areas where actions should be prioritised for reasons such as population density, existing health or environmental risks, pollution or contamination of water bodies, etc.

An important basis for successful implementation of the Urban Water Resources Management concept is the establishment of extensive monitoring platforms (urban and lake observatories) for the sources of water pollution as well as the recipient, Chao Lake. The observatories serve as an early warning system for operational water management (drinking water). Chapter 5 introduces a showcase of WP-C how such a monitoring system can be designed, deployed, and operated to inform stakeholders at the lake. It further shows how such a monitoring system can be elaborated by adding knowledge from modelling. The underlying strategy of this approach is that a thorough information about the current state of the lake system and its dynamics is helping drinking water suppliers and other stakeholders to optimize their process engineering, short-term reaction to water quality deterioration, and long-term planing. The large and shallow Chao Lake in the Anhui province serve as an example of tailored in situ monitoring of water quality dynamics (e.g. AWATOS buoy with multi-parameter probe (EXO2, YSI) for basic physical, chemical and biological variables and a multi-channel fluorescence probe (Phycoprobe, bbe moldaenke)) and modelling tools for water managers as well as the shallow Bao'an Lake in the Hubei province as a special case for the application of dynamic biomonitoring (e.g. Daphnia-Toximeter). In addition, a three-dimensional numerical model (GETM) of Chao Lake was set up to predict stratification dynamics as well as lake-wide circulation patterns and transport. Such information can be valuable for lake managers, e.g. for understanding large-scale wind-driven transport of scum-forming cyanobacteria, pollutant distribution or the occurrence of resuspension events. The research was supported by cooperation of Institute of Hydrobiology (TU Dresden), Department of Lake Research (UFZ Magdeburg), bbe Moldaenke GmbH, Nanjing Institute of Geography & Limnology (CAS) and Institute of Hydrobiology (CAS).

An Environmental Information System (EIS) provides data for the UWRM-concept, including required data infrastructures, interoperable simulation tools and web services. On the one hand, the combination of monitoring and modelling platforms in the EIS enables the identification of pollutant sources and pathways in the entire catchment area and, on the other hand, it is an important instrument for operational water management and long-term water quality forecasts. WP-D (Chap. 6) demonstrates a WebGIS approach for an environmental information system for time-series data from observation sites in Chaohu City (e.g. buoy, biomonitoring, Urban Laboratory, Urban Monitoring) to monitor water quality and hydrological parameters. in the second part of the chapter, a virtual geographic environment is introduced to present all available data sets for the Chao Lake catchment in one geographic context, such that the relevance and information for each data set are adequately visualised and the interaction with other data sets is coherently illustrated. Using the OpenGeoSys-DataExplorer (open source development within the project), all available data and models can be integrated and visualized in a geographical context. This work was carried out in cooperation of AMC, WISUTEC and the Department of Environmental Informatics (UFZ Leipzig).

The working group WP-E in collaboration with Institute of Groundwater Management (TU Dresden) and Department of Environmental Informatics (UFZ Leipzig) dealt in detail with the development of a groundwater model for the hydrological catchment area of Chao Lake (Chap. 7) which considers the water quality and water quantity. The groundwater model has been created containing the complete hydraulic system of the urban and suburban area including the network of rivers, streams and ditches. In addition, the model considered nutrient inputs from agricultural use in the Chao Lake area as well as other nutrient inputs via surface water that have an impact on groundwater.

1.5 Study Area

1.5.1 Chao Lake

Chao Lake (Chinese name: Chao Hu), known as Juchaohu Lake, also nicknamed Jiaohu Lake, is named after the lake shape likes a bird's nest. It is the largest lake in Anhui Province and one of the five major freshwater lakes in China (Fig. 1.1).

It is located in the central part of Anhui Province and on the left bank of the lower reaches of the Yangtze River valley with a longitude of 117°16′54″, 117°51′46″ and latitude 31°43′28″, 31°25′28″.

It belongs to the Yangtze River system and is a typical fault-depressed lake developed on the basis of structural basins. The total area of Lake Chao Basin is 13,350 km^2, and the area above Chao Lake is 9131 km^2, of which, the area of hills

Fig. 1.1 Chao Lake (Source: Marieke Frassl)

is 7735 km² (accounting for 84.7%), the polder area is 612 km² (6.7%) and the lake surface is 783 km² (accounting for 8.6%, when the water level is 12.00 m). The area below Chaohu Gate is 4219 km².

Chao Lake basin covers Hefei city, and Feidong, Feixi, Juchao, Hanshan, Hexian, Lujiang, Wuwei and Shucheng county. Chao Lake area extends across Baohe District, Feixi County, Feidong County, Chaohu City, Juchao District, Lujiang County.

Chao Lake is an important wetland in People's Republic of China (PRC), a national key scenic spot, and an important source of drinking water in Hefei and Chaohu. It posses multi-functions such as industrial water use, agricultural irrigation, flood control, fishery and tourism. It plays an extremely important role in the social and economic development of the basin.

In 2015, the annual gross domestic product (GDP) of the basin reached 692.805 billion Chinese yuan, accounting for 31.5% of the total GDP of Anhui Province, indicating that the basin is an important area and the central region affecting the stability of Anhui's economy and society.

The bottom of the lake is flat, with elevation varying from 5 to 10 m and the lowest 4.61 m. The lake basin slopes from northwest to southeast with an average slope of 0.96%. The lake can be divided into East half lake and West half lake along the line of Zhongmiao - Laoshan Island - Miaozuizi (here is the narrowest point of the Chao Lake with a width of 7.5 km). The water surface area of the West half lake is about 248 km² with the bottom elevation above 5.5 m; the surface water area of the East half lake is about 531 km² with the bottom elevation about 5 m. When the lake is at the multi-year-averaged water level of 8.37 m, the lake is 61.7 km long and 12.47 km wide with a water surface of 769.55 km², an average water depth of 2.89 m, a maximum water depth of 3.76 m, a volume of 2.07 billion m³ and a lake shore line of 155.7 km.

The lake basin is of subtropical humid monsoon climate. The annual mean temperature at the typical Zhongmiao Meteorological Station in the lake area is 16.1 °C, with the highest monthly averaged temperature 28.7 °C on July and the lowest monthly averaged temperature 2.8 °C on January. The average annual frost-free period is 263 days. Average annual rainfall is about 980 mm with a maximum annual rainfall of 1881.4 mm in 1991, a minimum annual rainfall of 516.7 mm in 1978. Thirty percent of the average annual precipitation distributes in spring, 48% in summer and 22% in autumn and winter.

The average water resources in the Chao Lake catchment area was 3.767 billion m³. Water resources was unevenly distributed during the year, 61.7% in May–September and 38.3% in the period of April to April. The interannual variation of the water resources was large. For example, in 1991 and 1969, the water resources were 9.74 and 7.19 billion m³ respectively, while it was only 1.35 billion m³ in the special dry year 1978.

There are 33 rivers and streams drain into Chao Lake. The main rivers are Hangbu River-Fengle River, Paihe River, Shiwuli River, Nanfei River, Dianbu River, Zhegao River, Zhaohe River and Baishitian River. Hangbu River-Fengle River, Paihe River, Nanfei River and Baishitian River account for more than 90% of the runoff. Hangbu River-Fengle River is the river with the largest amount of water injected into Chao

1 Introduction

Lake, followed by Nanfei River and Baishitian River, accounting for 65.1, 10.9 and 9.4% of the total runoff respectively. These rivers are most from the hills with large catchment area, shorter flow routine, steeper slope, fast confluence through polder area around the lake, into the Chao Lake. After collecting water from the south, west and north sides, Chao Lake gives outflows into the Yangtze River through the Yuxi River which lies to the southeast to Yuxi port of Wuwei County.

Chao Lake water level usually has a steep rise in flood season, with annual amplitude of 1.44–4.92 m. The average annual runoff into Chao Lake is about 4.25 billion m^3 with the maximum runoff of 10.88 billion m^3 into Chao Lake in 1991. The lake water transparency is generally 0.15–0.25 m, the corresponding water was brown. The water in the northeastern is more transparent, while the water in the western, northwestern and southern parts of the lake, as well as near the banks and estuaries severely affected by storm waves, are less transparent.

Lake eutrophication and algal blooms have always been an important environmental issue facing Chao Lake (Fig. 1.2).

The average total phosphorus and total nitrogen concentrations in Chao Lake in 2016 were 0.100 mg/l (greater than 0.099 mg/l in 2011) and 1.696 mg/l (higher than 1.14 mg/l in 2011), respectively. Those are the major pollutants of the lake. The water quality of all 9 measuring points surpassed Grade III standard of surface water. The maximum value of total nitrogen in the eastern half lake of the lake was 1.81 mg/l and the annual average was 1.30 mg/l (higher than 0.988 mg/l in 2011); the maximum total phosphorus was 0.101 mg/l, with an average annual value of 0.076 mg/l (less

Fig. 1.2 Algae bloom in Chao Lake (Source: Olaf Kolditz)

than 0.092 mg/l in 2011). The east half of the lake was slightly polluted with its water quality of IV. Among the 6 monitoring sites in the half lake, the water quality was of IV at 5 measuring sites and of V at the other one site. The western half of the lake was moderately polluted, with a max total nitrogen of 2.56 mg/l, an annual average of 2.20 mg/l (lower than 2.48 mg/l in 2011) and a max phosphorus of 0.122 mg/l, an annual average of 0.111 mg/l (lower than 0.114 mg/l in 2011). There is three measuring sites in the half lake. Its water quality was of IV at one sites and of V at the other two sites. From the view of monthly changes in 2016, the total phosphorus and total nitrogen concentrations in the whole lake were 0.091 mg/l and 2.925 mg/l in April respectively. After then total nitrogen concentration and total phosphorus concentration decreased, but rebounded from late May to June with the ratio of N to P in the lake decreasing as the bloom extended to the whole lake and the scale of the bloom increased. The average concentration of total phosphorus and total nitrogen in the whole lake dropped sharply because of the continued heavy rainfall from July to early August. The average concentration of total phosphorus and the total nitrogen in July were 21.5 and 5.8% of the previous month respectively. After the middle of August, the total nitrogen concentration of the whole lake remained stable, and the total phosphorus concentration increased rapidly. The total phosphorus concentration of the whole lake increased by 24.7 and 61.5% in August and September, respectively. In October, with the decrease of total phosphorus concentration and the increase of total nitrogen concentration, the ratio of nitrogen to phosphorus in the lake area increased. In the whole and the east half of the lake, the inferior V water quality were mainly concentrated in August and September, while in the west half lake those kind of water quality were mainly concentrated in April and also in August and September, were the main pollution indicators are pH, TP, TN.

In the 1950s, the main specie of the summer algal bloom in Chao Lake was Microcystis of cyanobacteria, with a small amount of Anabaena. In the spring and summer of 1963, algal blooms were dominated by Microcystis and Anabaena. By the 1980s, Microcystis still accounted for an absolute advantage (the whole lake average accounted for about 95%). Anabaena increased in winter and spring, but only accounting for 20%, less than 1% in summer and autumn. Although Microcystis is the dominant species of cyanobacteria in Chao Lake from the point of the occurrence frequency since 2010, Anabaena is the first dominant species from the point of the annual biomass, accounting for 54% of the total biomass of cyanobacteria, with Aphanizomenon accounting for 28%, Microcystis accounting for 17%. Microcystis was dominant in April and May, with a dominance of the whole lake in the period from June to September, especially in August. Aphanizomenon dominated most of the months before Microcystis dominance in June and after Microcystis bloom subsided in October, November. Anabaena appeared earlier, faded later than Aphanizomenon, from March to May in spring and from October to December in autumn and winter. However, Anabaena sp. occupied a considerable proportion of eastern half lake at certain sites in summer. The population of cyanobacteria in Chao Lake possessed distinct seasonal succession: Anabaena from spring - Aphanizomenon at spring and summer - Microcystis in summer - Aphanizomenon in autumn - Anabaena in winter. Cyanobacteria in spring and winter were higher in the western lake area of Chao

1 Introduction

Lake, especially in the mouths of rivers flowing into the lake. The algae usually clusters to the north off-shore area under the influence of south wind and resulting in a sharp increase in the number of algae. When encountering a west wind, cyanobacteria bloom occurs in the water source area of Chaohu City, which is located at the water outlet of the northeast of Chao Lake, under west wind field and throughput water current. From April to mid-May, 2016, the density of algae in Chao Lake was generally stable, and the density of algae in the west half lake and eastern half lake of the lake remained relatively low, basically below 2 million cells/l. From mid-May to mid-August, the highest concentration in the western and the eastern reached 8.94 and 8.17 million cells/l. From mid-August to early October, the density of the algae in Chao Lake fluctuated violently and the density of algae reached one year's high value, the maximum algae density of the west and the east reached 24.24 and 11.54 million cells/l, respectively. At the time, the algae density in the western was significantly higher than that of the eastern. In the late of October, the density of algae in the Chao Lake and the eastern rapidly dropped below 2 million cells/l. At Bakou and Chaohu Shipyard measuring points, which indicating status of drinking water sources quality, algae density ranged from 5 to 25 million cells/l in 2016, with a mean average density of algae of 1.24 and 2.18 million cells/l, and the annual algal density fluctuations significantly lower than the average algal density in the eastern and the western semi-lakes. Algal at these two sites only exceeded 10 million/l on May 24 and July 26. From April to October of 2016, the degree of bloom in the whole lake, the western half lake and the eastern half lake were all from no obvious bloom to mild bloom. No obvious algae bloom accounted for 37.2%, slight algae bloom 51.0%, mild algae bloom 11.8%. Eighty-eight times the algae bloom monitoring was effectively performed by satellite remote sensing technology in the Chao Lake in 2016. Of which, 24 were absent of obvious bloom, accounting for 27.3%, 55 were of sporadic bloom, accounting for 62.4% and 7 were of particle water bloom, accounting for 8.0%; 2 were of a regional bloom, accounting for 2.3%. The largest scale cyanobacterial blooms were detected on June 26 with an area of about 237.6 km^2, accounting for 31.2% of the whole area Chao Lake. The occurrence of cyanobacteria blooms mainly occurred in the last ten-day of June and September and at the beginning of October. There were more frequent sporadic blooms from July to August.

1.5.2 Chaohu City

Chaohu city is located in the middle of Anhui Province, where close to Chao Lake. The city has 2063 km^2 of area with 940 thousand population and nearly 400 thousand permanent resident population. The average rainfall around the city is about 1000–1158 mm, and about 65% of rainfall concentrated during June to August every year (Zhang and Liao 2016).

The inland river in the city is a complex network system, which constructed by Huancheng River, Xier Pool, Lujia River and so on, as it shows in Fig. 1.3 and Table 1.1.

According to the picture, the main river cross includes Guishan-Xipie-Shuangqiao River system, West Huancheng - East Huancheng River system, Yuxi-Tian River system, Xiaowangzhuang-Xier-Lujia River system and Baoshu-Juzhang River system. Tian River was the main river cross of Yuxi River. Two dams were set on both of the up and down stream of Tian River for flood protection, also made the river closed from outside river system. The West-East Huancheng river system goes around the centre of the city with 4.4 km length and 0.16 km^2 area, and connected to the Chao Lake. The area of Xier Pool is 0.038 km^2, which collects the water comes from

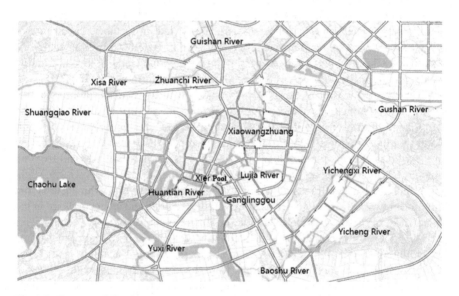

Fig. 1.3 The map of Chaohu city

Table 1.1 The basic information of Inland River

River system	River	Length (km)	Width (m)
Guishan-Xipie-Shuangqiao	Guishan River	~3	10–15
	Xiepie River	3.3	20–50
	Shuangqiao River	1.3	20–40
West Huancheng - East Huancheng	West Huancheng River	2.7	40
	Zuanchi River	2.0	Silt up
	East Huancheng River	1.5	40
Xiaowangzhuang-Xier-Lujia	Xiaowangzhuang River	2	4
	Xier Pool	0.038 km^2	
	Lujia River	1.5	40
Yuxi-Tian	Tian River	1.5	80
Baoshu-Juzhang	Anchengyu River	2.5	30

1 Introduction

Xiaowangzhuang River, and connected to the East Huangcheng River and Tian river. Both the Lujia River and Xier Lake are blocked from outside river system as the silt up happened for many years.

1.5.2.1 The Investigation of Pollution Source in Chaohu City

It is possible to find out the distribution of pollution source in Chaohu city by combining tracking and tracing analysis. The research area (Zhou and Liao 2014) including the old and the new district of Chaohu city, which are residential area in the city with both combined sewer systems and separated sewer systems. There are two sewage pumping station, three rain and sewage combined pumping stations and one wastewater treatment plants in the research area. The area is marked in Fig. 1.4 by a green polygon.

(1) **Water quantity data**

The annual water consumption data in the research area, which was come from 89442 houses and grouped by neighbourhood, was collected during 2013 (Zhou and Liao 2014). The water consumption data of pollution source can also be obtained after combining both the neighbourhood water consumption data and pollution source data (shown in Table 1.2 and Fig. 1.5). As there is no such industry which need huge quantity of water (for example, beverage industry and landscape architecture) in the

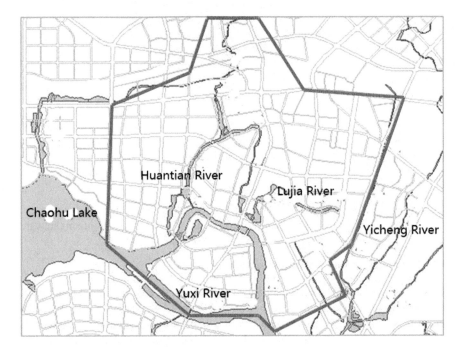

Fig. 1.4 The research area

Table 1.2 The water consumption data of each pollution source

Pollution source	Source number	Annual consumption (1000 tons)
Industry	5	1682
Resident	255	6293
Public services	466	5629
Catering	8	37
Hospital	18	838
Total	752	14,479

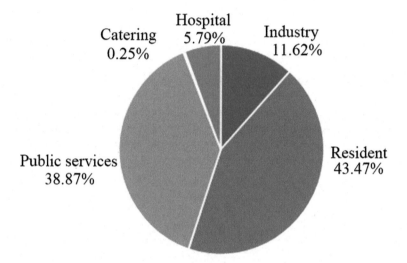

Fig. 1.5 The water consumption data of each pollution source

research area, the waste water quantity can be approached by 90% of consumption water (Zhou and Liao 2014).

(2) **Water quality data**

Considering the pollution source data, pipe network data and the water quantity data from both pumping stations and waste water plant, it is possible to confirm the water quality of each pollution source (Table 1.3) (Zhou and Liao 2014).

(3) **The distribution of pollution source**

Based on the tracking and tracing analysis, the distribution of pollution sources can be found in the research area. According to the waste water plant and the pipeline network, the analysis is able to figure out the distribution of pollution sources in the area which are illustrated in Fig. 1.6 (The green point is the waste water plant, and the red points are the location of pollution source).

Table 1.3 The water quality of each pollution source

Pollution source	COD (mg/l)	NH3-N (mg/l)	TN (mg/l)	TP (mg/l)	SS (mg/l)
Industry	500	40	60	6	150
Resident	358.46	48.55	59.42	4.17	46.62
Public services	358.46	48.55	59.42	4.17	46.62
Catering	1560	3.15	37.25	6.22	252
Hospital	256.48	20.83	28.24	2.16	40

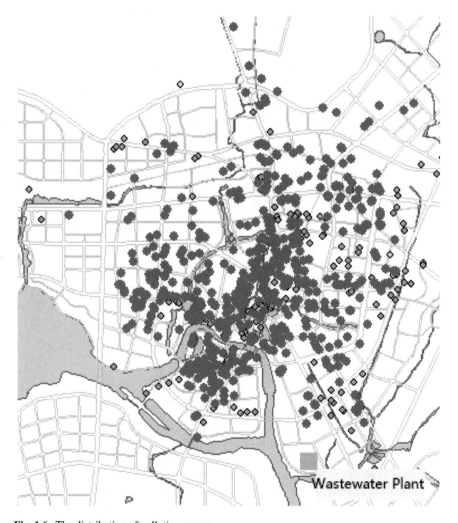

Fig. 1.6 The distribution of pollution source

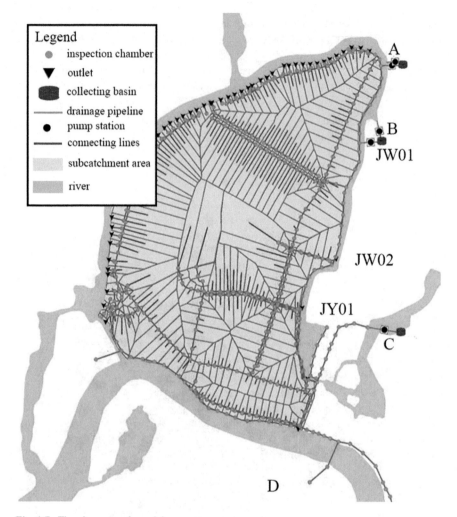

Fig. 1.7 The pipe network model

1.5.2.2 The Pipe Network Model of Chaohu City

The pipe network model was built based on the collected data (Xie et al. 2017). The MIKE URBAN, a software used for pipe network model simulation (Brown 1976; Xie et al. 2016) was used as the tool to set up the model. Also, the ENVI5.1 model and ENVI and the GIS were used on the modelling process (Xie et al. 2017). The generalisation of the pipe network system was based on the data of network, pollution, pumping stations, the type of land using and terrain. The details are given as follow:

1 Introduction

(1) Drainage pipe network

The model of drainage pipe network was built by ArcGIS10.1, which including sewage pipes, storm sewer, combined pipes, manholes. The interface of MIKE URBAN was used to input the ArcGIS data into system and set up the structure of the model.

(2) Sub-catchment area

The watershed boundary was found by ArcGIS10.1, and the catchment areas were set based on the boundary. Then the sub-catchment areas were confirmed by applying Theissen polygon principle on each manholes. There are 234 sub-catchment areas in total after the process.

(3) The type of land using

The classification of the type of land using was bound to each sub-catchment areas in the model by ArcGIS10.1 (Xie et al. 2017).

(4) Pumping station

The four pumping stations were treated as boundary condition of the model, and each of them was set at the nearest manhole.

(5) Pollution source

The pollution sources were connected to the manhole and pipe network. The discharge of each pollution source were decided by the water consumption data. There are 125 pollution sources in the area (Xie 2017).

After all the process were finished, the pipe network model was built as Fig. 1.7.

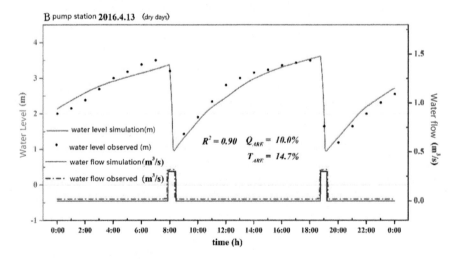

Fig. 1.8 The result of model on pumping station B on the dry day

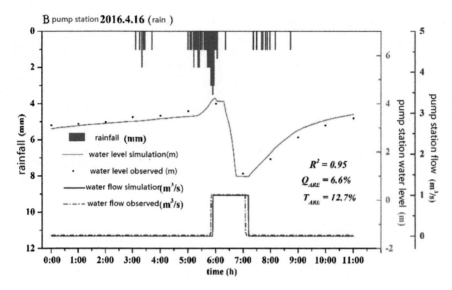

Fig. 1.9 The result of model on pumping station B on the rainy day

The boundary conditions of the model were designed by rainfall data. The time step of the runoff simulation was set as 1 min according to the rainfall series, and the time step of pipe network simulation was set between 1 and 5 seconds for the stabilisation requirement. The model was used to simulate the situation in both a dry day (0:00 to 24:00 May 13rd, 2016) and a rainy day (May 16th, 2016). The water level data of the pumping station B was collect to compare with the result of model simulation (Figs. 1.8 and 1.9). The dry day (8:00 to 22:00, August 10th, 2014) water quality data of the manhole JW01, which is the first manhole behind station B, was collected to compare with the simulation results of model (Fig. 1.10). And the rainy day (June 2nd, 2015) water quality data of the manhole JY02 was also collected for the comparing. Both of the data and the results of model simulations were given as Fig. 1.11 (Xie 2017).

1.5.2.3 The Inland River Model of Chaohu City

The inland river model was built based on the data of river direction, river length and hydraulic structure by MIKE11 (DHI, MIKE11 2015; Zhang and Liao 2016). The data of 150 river sections are also collected for model developing. The generalization of the model including 6 river courses, 5 boundary points, 13 water gates (the small

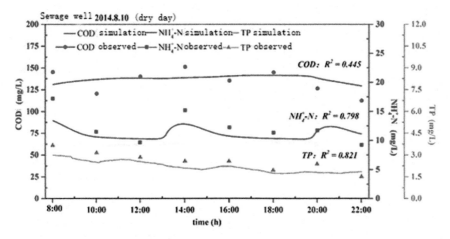

Fig. 1.10 The result of model manhole on the dry day

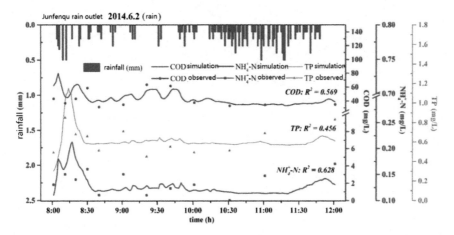

Fig. 1.11 The result of model on manhole JY02 on the rainy day

blocks on the map) and 1 pumping station. Each part of the river were labelled by the codes DGH, XHCH, DHCH, TH, XEC, LJH on the map. The numbers following the code are the length of the point on the river (the unit is meter). All these feature are given in Fig. 1.12.

We also considered the following during the modelling:

(1) **The time step of model**

Considering the stabilisation and the time costing, the time step of the model was 5 s, and the space step was decided by MIKE11 based on the river sections.

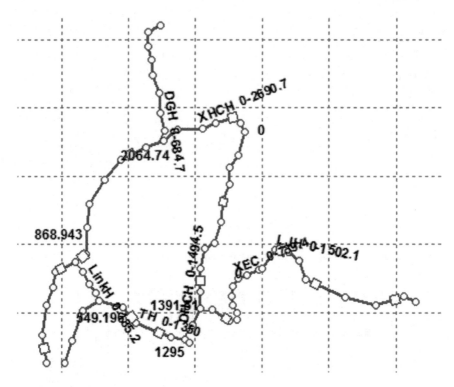

Fig. 1.12 The generalization of model

(2) **The boundary condition**

The data of the dry days (November 28th and December 10th, 2013) and the rainy day (June 2015) were collected as boundary condition with frequency once an hour. The water quality data of each river sections on the same time were also collected (Zhang and Liao 2016).

(3) **The roughness rate of the river bed**

As one of the most important parameter of the model, the roughness rates (Zhang and Liao 2016) of the river bed were set in range 0.25–0.4 based on different and adjusted based on the data of water level and discharge.

After the process above, the model was used for validation. The hydrodynamic model was used to simulate the water level and discharge on December 10th, 2013, and the water quality model was used to simulate the water quality from December 5th to December 10th, 2013. The results of 4 points on the river (Xishenggong Bridge, Renmin Bridge, Tuanjie Bridge and Yangba Bridge) are given as Figs. 1.13 and 1.14. Based on the results, both of the hydro-dynamic model and water quality model provided a superior simulation.

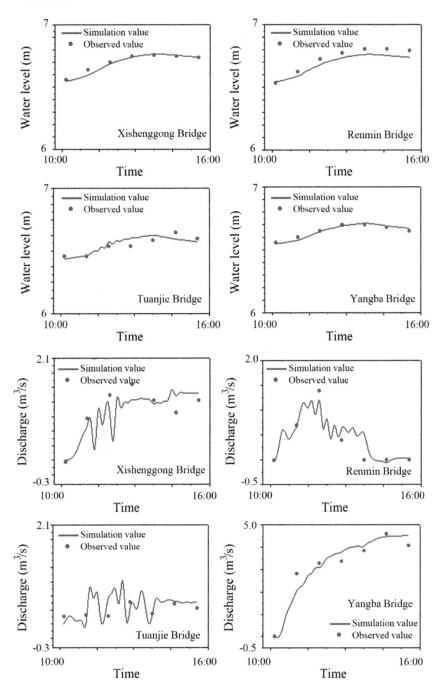

Fig. 1.13 The result of hydrodynamic model

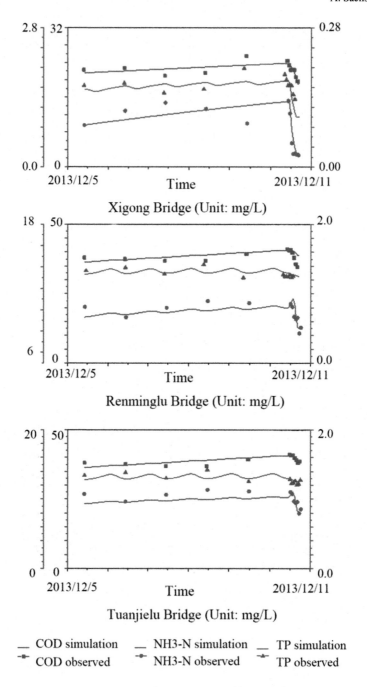

Fig. 1.14 The result of water quality model

1.5.2.4 The Water Environment Information System in Chaohu City

All the data, models were combined into an information system at the end of the research (Hu and Liao 2017). The system was built on Visual Studio 2010 with Microsoft Access 2003 and ArcGIS Engine10.1. There are mainly 3 basic function modules: Map operation module, model results case module and staff and equipment information module. The Map operation module was shown as Fig. 1.15. In this part, the ArcGIS Engine 10.1 bound with Visual Studio 2010 by C# to achieve basic GIS function such as map roaming, bookmark, feature selection and GIS information searching. Model results case module searches the result of the inland river model above and visualizes the selected data. In this part, the results can be searched by time and location and displayed by MSchart, a controller in Visual Studio for data visualization. Also, the results of models were bound to the ArcGIS information, which means it is possible to search the model results through the map roaming and feature selection. The windows of this module were shown as Figs. 1.16 and 1.17. The staff and equipment information module is based on SQL Server and OLEdb method (a method used in Visual Studio for data operation) and used for staff & equipment information searching.

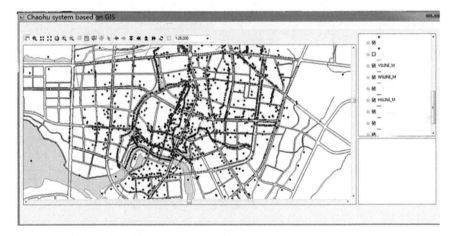

Fig. 1.15 The Map operation module of the system

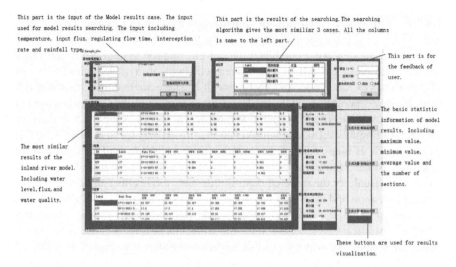

Fig. 1.16 The search function of Model results case module

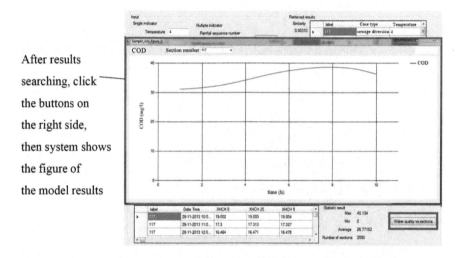

Fig. 1.17 The results visualization function of Model results case module

1.6 Literature Research - State of the Art

Due to the rapid economic development in the last decades, China is now facing serious environmental challenges (Wohlfart et al. 2016; Yue et al. 2015), including water pollution, water shortage, floods, water salinisation, groundwater depletion, catchment soil erosion, ecosystem deterioration and biodiversity loss. Especially water pollution is a large risk to the Chinese society today. More than 40% of China's rivers are severely polluted, and 80% of its lakes suffer from eutrophication (Liu and

Yang 2012). Since most of the drinking water in China is collected from surface water, harmful algal blooms and chemical spills in rivers or lakes are of great concern (Chen and Liu 2014; Tang et al. 2006; Wang et al. 2008). Wells and aquifers are contaminated with fertilisers, pesticide residues and heavy metals from mining (Li et al. 2013b) and petrochemical industries (Li et al. 2013a). About 300 million people in rural areas are lacking safe drinking water resources (Dou et al. 2014; Liu and Yang 2012). As a result, water resources management and wastewater treatment coupled with pollution control policies are urgently needed to improve the water environment as a whole. Therefore, China has launched a number of water pollution control plans over the last decade, to deal with the existing problems (Chinese Ministry of Environmental Protection 2016). Water has become a major topic within the 11th Chinese Five-Year Plan and the "Major Science and Technology Program for Water Pollution Control and Treatment" (Mega Water Projects 2006–2020) has been initiated by the Chinese government. It's goals are to improve wastewater treatment technology and drinking water supply as well as to rehabilitate polluted rivers and lakes (Chen et al. 2015a, b; Zhi et al. 2016).

There is a wealth of scientific literature on various aspects of water quality, algal blooms, accumulation of pollutants in lake sediments, and the ecological status of Chao Lake (187 ISI publications as of 19 August 2013). Little research has been conducted on ways to improve the lake's water quality. Many factors contribute to the lake's extremely high pollutant level, including nutrient inflow, such as eutrophication (Shang and Shang 2007), phosphate accumulation in sediments (Zan et al. 2011, 2012) or the formation of cyanotoxins (Chen et al. 2008), and contamination by heavy metals (Han et al. 2011) and organic toxins (He et al. 2012).

1.6.1 Most Cited Literature Around Chao Lake: 1990–2017

Numerous studies have intensively analysed Chao Lake and its catchment: Chao Lake, with an average depth of about 3 m, is one of the five largest fresh-water lakes in China (Figs. 1.18 and 1.19). Before the 1950s, it was well-known for its scenic beauty land for its richness in aquatic products. Over the last decades, with the population growth and economic development in the drainage area, nutrient-rich pollutants have drained increasingly into the lake (Kang et al. 2016b). Especially the Nanfei River is by many measures the most heavily polluted tributary to Chao Lake with high concentrations of persistent organic pollutants in suspended particulate matter (Liu et al. 2017). But the nevertheless the pollution status of the most of the lake's tributaries and estuaries is still unknown (Wang et al. 2016a). This pollution has resulted in lake eutrophication, lake ecosystem structure disharmony and lake ecosystem damage (Yang et al. 2016a). The historical heavy eutrophication in Chao Lake is the main source of increasing organic matter sediment (algae and terrestrial input) in the eastern part of the lake (Xu et al. 2017). High eutrophication levels occurred in the western Chao Lake in spring and summer, whereas high levels occurred in the eastern lake, especially in the middle of the lake, in autumn and winter (Yang et al. 2016a).

Fig. 1.18 Overview of publications regarding Chao Lake, limited to countries

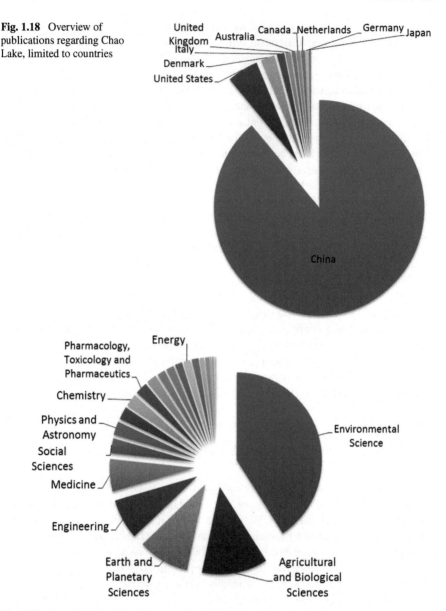

Fig. 1.19 Overview of publications regarding Chao Lake, limited to subjects

The increasing pollution of freshwater lakes in China, especially the Chao Lake dominates the literature. The most frequently cited topic around the Chao Lake deals with microcystin (cited more than 200 times). Microcystin (MC) is produced by cyanobacteria in freshwater (Chen et al. 2005).

Blooms of cyanobacteria developed during summer in Chao Lake (Li et al. 2016c). An intensive bloom was observed in 1980 (Kong et al. 2017). Chao Lake has been experiencing lake-wide toxic Microcystis blooms in recent decades. The abundances of total and toxic Microcystis and Microcystin concentrations showed significant positive correlation with the total phosphorus and water temperature, suggesting that increases in temperature together with the phosphorus concentrations may promote more frequent toxic Microcystis blooms and higher concentrations of Microcystin (Yu et al. 2014).

Human intoxications by hepatotoxic Microcystins (MCs) can lead to liver damage or even death. Some reports suggest that the incidence of human primary liver cancer in the eastern region of China is related to the presence of Microcystins found in drinking water. Microcystins could accumulate in living organisms and transfer through the food chain, consequently threatening human health, when irrigation with contaminative lake water containing a variety of Microcystins (Chen et al. 2009; Jin and Chang 2013). According to the study by Chen et al. (2009) fishermen at Chao Lake, are exposed to very high levels of Microcystin contamination. So far, there have been no direct evidences of MC occurrence in human tissue in consequence of exposure to MC. In this study, the autors improved cleanup procedures for detecting MCs in serum sample using liquid chromatography-mass spectrometry.

Most of the aquatic products from the three large lakes in China seem to be unsafe for human consumption due to Microcystin accumulations, with the estimated daily intake (EDI) values 5–148 times, 2–50 times and 1.5–4 times higher than the tolerable daily intake (TDI) value in Taihu, Chao Lake and Dianchi, respectively. In addition, the toxin accumulation in the harvested organisms varied intensity from month to month and by species which suggests that consumption risks may be reduced or avoided by either adjusting the legal fishing seasons or the species of fish and shellfish harvested (Peng et al. 2010). Xie et al. (2005) as well as (Jiang et al. 2017) describes the bioaccumulation of hepatotoxic Microcystins in freshwater fishes from the large, shallow, eutrophic Chao Lake in September 2003, when there were heavy surface blooms of toxic cyanobacteria. The MCs showed a tendency to accumulate up the food chain, and fish at the top of the food chain were at high risk of exposure to MCs in Chao Lake.The study showed that one hundred grams of fish muscle would contain about 1.3–25 times the recommended tolerable daily intake of Microcystin by humans, indicating that fish are already severely contaminated by MCs and that the local authorities should warn the public of the risk of poisoning by eating the contaminated fish (Xie et al. 2005). Microcystins, which are the secondary metabolites of cyanobacteria, may accumulate in fish via feeding or bioaccumulation, and finally harm human health through the food chain. Protein is the most expensive component in fish feeds. Based on cost effectiveness, availability, and crude protein content, cyanobacteria seem to have considerable potential in fish feeds (Dong et al. 2012). Shrimp are also affected by Microcystin contamination (Chen and Xie 2005). Among the shrimp muscle samples analyzed, 31% were above the provisional WHO TDI level, suggesting the risk of consuming shrimps in Chao Lake. It is recommended that edible mussels should not be collected for human consumption during toxic cyanobacterial blooms in Chao Lake (Chen and Xie 2008).

One possible method in the elimination of cyanobacteria blooms in the lake is the increased colonization with carp. The results of the study by Xie and Liu (2001) suggest the applicability of a new food-web manipulation (increased stocking with filter-feeding fish) for controlling cyanobacteria blooms in hypereutrophic lakes, especially in lakes where nutrient inputs cannot be reduced sufficiently, and where zooplankton cannot effectively control phytoplankton production. Also, the elimination of overwintering cyanobacteria in sediment is vital to control cyanobacterial blooms. Therefore the sediment plow-tillage method was introduced as an innovative technique for eliminating overwintering cyanobacteria in sediments from Chao Lake (Zhou et al. 2016). A bacterium capable of degrading Microcystin was tentatively identified as Paucibacter sp. The optimum temperature and initial pH for MC degradation were 25–30 °C and pH 6–9, respectively (You et al. 2014).

In Chao Lake a collapse of the food web toward a simplified structure and decreasing biodiversity and trophic interactions can be observed. The lake ecosystem was approaching an immature but stable status from the 1950s to the 2000s. Discussion about the potential driving factors and underlying mechanisms, hypothesizing that hydrological regulation may play a significant role in driving all of these changes in Chao Lake in addition to eutrophication and intensive fishery. As a new new management tool, Ecopath, a typical steady state model for trophic mass-balance analysis in ecosystems can be considered (Jørgensen 2016; Kong et al. 2016).

Accumulation and degradation of cyanobacteria in water cause peak values in taste and odor (T&O) compounds in western Chao Lake which leads to exceeding odor thresholds (Jiang et al. 2016).

The extension and frequency of algal blooms like cyanobacteria in surface waters can be monitored using remote sensing techniques and in-situ monitoring. Knowledge of their vertical distribution is fundamental to determine total phytoplankton biomass in the shallow eutrophic lake of Chao Lake as well as for forecasting algae bloom (Bo et al. 2016; Xue et al. 2017). The estimates of total phytoplankton biomass were both consistent with in situ measurements and consistent for remotely sensed reflectance observations made on the same day and on consecutive days (Li et al. 2017). Data from satellite images are a good tool for analysing the distribution of cyanobacteria in Chao Lake. The cyanobacteria in Chao Lake were dominated by species of Microcystis and Anabaena. Microcystis reached its peak in June, and Anabaena had peaks in May and November, with an overall biomass that was higher than that of Microcystis (Zhang et al. 2016c).

Lake eutrophication in general has become a serious environmental problem in China. Current status and future tendency of lake eutrophication in China was analysed by Jin et al. (2005).

The agricultural intensification and the extensive use of fertilizers leads to a heavy metal pollution in sediments in Chao Lake Valley (Tang et al. 2010). The sediment in the drinking water source area (DWSA) of the Chao Lake was threatened by heavy metals from other areas (He et al. 2016a). Surface sediment samples were analysed to determine the concentrations of 25 metal elements. The results of the pollution indices indicate that Chao Lake was weakly to moderately affected by Ti, V, Cr, Mn, Co, and Ni but was severely contaminated by Hg and Cd. The overall pollution level in the eastern lake was higher than that in the western lake with respect to the pollution level index (PLI). The contribution from industrial and municipal impact was negligible, despite the rapid urbanization around the studied area (Wang et al. 2016b). The Nanfei River (Anhui Province, China) is a severely polluted urban river that flows into Chao Lake. Multivariate statistical analysis demonstrated that Hg, Cu, Cr, Cd, and Ni may have originated from industrial activities, whereas As and Pb came from agricultural activities. Therefore, sediments in the Nanfei River watershed are heavily polluted and urgent measures should be taken to remedy the status (Shiguang et al. 2016).

Nevertheless the knowledge on the distribution of Total arsenic (As) between the particle and aqueous phases in freshwater lakes remains largely unknown. The high proportions of Algae, especially Cyanophyta in the composition of suspended particulate matter in the three largest shallow lakes of China (e.g. Chao Lake), might play an important role in affecting the As distribution between the aqueous and particulate phases in aquatic ecosystem (Yang et al. 2016b).

Seventeen polycyclic aromatic hydrocarbon (PAH) compounds were determined in surface sediments collected from the Chao Lake. The sites, located in the downstream of the local iron-steel manufacturing plant located in the Nanfei River contained very high concentrations of 17 PAH. Five factors were identified to generate PAH, including industrial waste, wood/biomass burning, diagenetic origin, domestic coal combustion, and industrial combustion (Han et al. 2016; He et al. 2016b; Li et al. 2016a; Zhang et al. 2016a).

The risk assessment on phosphorus release from stream-bed sediments has its maximum content emerged in autumn, followed by winter, and the minimum occurred in spring or summer. The negative and significant correlations of water total phosphorus and sedimentary total phosphorus may indicate that the risk of sedimentary phosphorus release was great in the western and eastern Chao Lake during algae bloom sedimentation (Li et al. 2016b; Liu et al. 2016b; Wu et al. 2016).

China is the most polluted region as affected by Tetrabromobisphenol A (TBBPA) compared with other countries. Tetrabromobisphenol A (TBBPA), a currently intensively used brominated flame retardant, is employed primarily as a reactive flame retardant in printed circuit boards but also has additive applications in several types of polymers. Sources of TBBPA in China are mainly derived from the primitive e-waste dismantling, TBBPA manufacturing and processing of TBBPA-based materials and can be found for example in Chao Lake, Anhui (industry concentration site) (Liu et al. 2016a).

The residual levels of DDT-related contaminants (DDX) were higher in the surface and core sediments in the western lake area than in other lake areas, which might be due to the combined inflow effects of municipal sewage, industrial wastewater and agricultural runoff. The DDX residues in the sediment cores reached peak values in the late 1970s or early 1980s (Kang et al. 2016a).

Environmental pollution caused by synthetic pyrethroid insecticides has received a great deal of attention with the increase in usage recently. The medication and insecticide permethrin was found in the water, air and sediment of the Chao Lake. Sediment was the largest sink. In order to protect 95% of species in Chao Lake, the maximum annual input amount of permethrin should be controlled below 78.2 $t\,a^{-1}$ (Liu et al. 2016d; Wang et al. 2016a). Aquatic biota have long been recognized as bioindicators of the contamination caused by hydrophobic organic contaminants (HOCs) in aquatic environments. Culter erythropterus and Protosalanx hyalocranius (fish) are the most sensitive to organochlorine pesticides (OCPs) in Chao Lake and can therefore serve as indicators of the lake's health and assist in the assessment of OCPs risks to human health (Liu et al. 2016c; Zhang et al. 2016b).

The composition and diversity of benthic macroinvertebrate communities in the five largest freshwater lakes of China was analysed by Cai et al. (2012). Benthic assemblages in Chao Lake changed greatly over two decades. These observed changes are discussed relative to the ability of individual different taxa to tolerate eutrophication, algal blooms, Microcystins and associated habitat deterioration (Cai et al. 2012).

Groundwater is the important source of water for drinking, washing, and irrigation in the watershed of Chao Lake (Yang et al. 2016c). A seasonal patterns of nitrate-N concentration is found in the groundwater close to vegetable land, orchard and rice field possibly due to different time and amount of fertilizer application (Qian et al. 2007; Wang et al. 2014; Yang et al. 2014). Cyanobacterial proliferation threatens the safety of drinking water supplies worldwide and in the drinking water source of Chao Lake (Shang et al. 2015). The groundwater near Chao Lake poses a significant health risk for the local residents when used for drinking water (Yang et al. 2016c). All collected groundwater samples (especially from the western coast of the lake) from the Chao Lake region had detectable concentrations of Microcystins. High concentrations of Microcystin observed in this lake are the result of the bloom's high proportion of toxic cyanobacteria strains (Yang et al. 2016c).

The majority of the Chao Lake Basin exhibits intermediate to high flood disaster resilience, but there are distinct variations within the basin. The resilience is clearly influenced by the natural dimension indexes, the highest resilience levels are mainly located in the hilly, mountainous regions, and the lowest resilience levels mainly occur in the south-southeast plain of the Chao Lake Basin and the river estuary of the Chao Lake (Sun et al. 2016). Some studies are available which estimates the value of ecosystem services such as material production, air purification, water conservation, biodiversity, recreation, species conservation, education and scientific research along Chao Lake (Li and Gao 2016). This is important because the provision of ecosystem services could be affected by landscape structural changes (Zhang and Gao 2016).

Effective lake management can help preserve the stability and long-term dynamics of a sensitive ecosystem such as the large shallow Chao Lake located in one of the most densely populated areas in China. Often, however, the interactions between water level control and nutrient loading are overlooked and largely underestimated (Kong et al. 2017).

Integrated engineering techniques had the potential to remediate heavily polluted rivers (Shuangqiao River, one of the most heavily polluted rivers inflowing to the Chao Lake) by a multi-pond constructed wetlands system and an in situ purification system consisted of sediment dredging, hydrophytes restoration and artificial floating islands (Fang et al. 2016; Weigen et al. 1990).

References

Bo X, Yanchuang Z, Xinyuan W, and Xin Z. Research on algae blooms forecasting based on the multivariate data driven method: a case study of the Chaohu Lake. IOP Conf. Ser.: Earth Environ. Sci. **46**(1), 012044 (2016)

Brown LA. National Symposium, on Urban Hydrology and Sediment Control, in *Proceedings: National Symposium on Urban Hydrology, Hydraulics, and Sediment Control, July 27–29, 1976*, vol. 111 (College of Engineering, University of Kentucky, ORES Publications, 1976), p. 99

Cai Y, Jiang J, Zhang L, Chen Y, and Gong Z. Simplification of macrozoobenthic assemblages related to anthropogenic eutrophication and cyanobacterial blooms in two large shallow subtropical lakes in China. Aquat. Ecosyst. Health Manage. **15**(1), 81–91 (2012). ISSN 1463-4988. https://doi.org/10.1080/14634988.2011.627017

Chen Y, and Liu QQ. On the horizontal distribution of algal-bloom in Chaohu Lake and its formation process. Acta Mech. Sin. **30**(5), 656–666 (2014)

Chen J, and Xie P. Tissue distributions and seasonal dynamics of the hepatotoxic microcystins-LR and -RR in two freshwater shrimps, Palaemon modestus and Macrobrachium nipponensis, from a large shallow, eutrophic lake of the subtropical China. Toxicon **45**(5), 615–625 (2005). https://doi.org/10.1016/j.toxicon.2005.01.003

Chen J, and Xie P. Accumulation of hepatotoxic microcystes in freshwater mussels, aquatic insect larvae and oligochaetes in a large, shallow eutrophic lake (Lake Chaohu) of subtropical China. Fresenius Environ. Bull. **17**(7 A), 849–854 (2008)

Chen J, Xie P, Guo LG, Zheng L, and Ni LY. Tissue distributions and seasonal dynamics of the hepatotoxic microcystins-LR and -RR in a freshwater snail (Bellamya aeruginosa) from a large shallow, eutrophic lake of the subtropical China. Environ. Pollut. **134**(3), 423–430 (2005). ISSN 0269-7491. https://doi.org/10.1016/j.envpol.2004.09.014

Chen J, and Xie P et al. Accumulation of hepatotoxic microcystins in freshwater mussels, aquatic insect larvae and oligochaetes in a large, shallow eutrophic lake (Lake Chaohu) of subtropical China. Fresenius Environ. Bull. **17**, 849–854 (2008)

Chen J, Xie P, Li L, and Xu J. First identification of the hepatotoxic microcystins in the serum of a chronically exposed human population together with indication of hepatocellular damage. Toxicol. Sci. **108**(1), 81–89 (2009). https://doi.org/10.1093/toxsci/kfp009

Chen C, Börnick H, Cai Q, Dai X, Jähnig SC, Kong Y, Krebs P, Kuenzer C, Kunstmann H, Liu Y, Nixdorf E, Pang Z, Rode M, Schueth C, Song Y, Yue T, Zhou K, Zhang J, and Kolditz O. Challenges and opportunities of German-Chinese cooperation in water science and technology. Environ. Earth Sci. **73**(8), 4861–4871 (2015a). ISSN 1866-6299. https://doi.org/10.1007/s12665-015-4149-5

Chen C, Sun F, and Kolditz O. Design and integration of a GIS-based data model for the regional hydrologic simulation in Meijiang watershed, China. Environ. Earth Sci. **74**(10), 7147–7158 (2015b). ISSN 1866-6299. https://doi.org/10.1007/s12665-015-4734-7

DHI, MIKE11. A Modelling System for Rivers and Channels-Reference Manual (DHI: Hørsholm, Denmark, 2015)

Dong G, Xie S, Zhu X, Han D, and Yang Y. Nutri-toxicological effects of cyanobacteria on fish. Shengtai Xuebao/ Acta Ecol. Sin. **32**(19), 6233–6241 (2012). https://doi.org/10.5846/stxb201109061310

Dou M, Wang Y, and Li C. Oil leak contaminates tap water: a view of drinking water security crisis in China. Environ. Earth Sci. **72**(10, SI), 4219–4221 (2014). ISSN 1866-6280. https://doi.org/10.1007/s12665-014-3556-3

Fang T, Bao S, Sima X, Jiang H, Zhu W, and Tang W. Study on the application of integrated eco-engineering in purifying eutrophic river waters. Ecol. Eng. **94**, 320–328 (2016). ISSN 0925-8574. https://doi.org/10.1016/j.ecoleng.2016.06.003

Han YM, Cao JJ, Kenna TC, Yan B, Jin ZD, Wu F, and An ZS. Distribution and ecotoxicological significance of trace element contamination in a 150 yr record of sediments in Lake Chaohu, eastern China. J. Environ. Monit. **13**(3), 743–752 (2011)

Han YM, Wei C, Huang R-J, Bandowe BAM, Ho SSH, Cao JJ, Jin ZD, Xu BQ, Gao SP, Tie XX, An ZS, and Wilcke W. Reconstruction of atmospheric soot history in inland regions from lake sediments over the past 150 years. Sci. Rep. **6**, 19151 (2016). https://doi.org/10.1038/srep19151

He W, Qin N, He Q-S, Wang Y, Kong XZ, and Xu F-L. Characterization, ecological and health risks of DDTs and HCHs in water from a large shallow Chinese lake. Ecol. Inform. **12**, 77–84 (2012)

He W, Bai Z-L, Liu W-X, Kong X-Z, Yang B, Yang C, Jorgensen SE, and Xu F-L. Occurrence, spatial distribution, sources, and risks of polychlorinated biphenyls and heavy metals in surface sediments from a large eutrophic Chinese lake (Lake Chaohu). Environ. Sci. Pollut. Res. **23**(11), 10335–10348 (2016a). ISSN 0944-1344. https://doi.org/10.1007/s11356-015-6001-6. International Conference on Contaminated Sediments (ContaSed-2015), Ascona, Switzerland, 08–13 Mar 2015

He W, Yang C, Liu W, He Q, Wang Q, Li Y, Kong X, Lan X, and Xu F. The partitioning behavior of persistent toxicant organic contaminants in eutrophic sediments: coefficients and effects of fluorescent organic matter and particle size. Environ. Pollut. **219**, 724–734 (2016b). ISSN 0269-7491. https://doi.org/10.1016/j.envpol.2016.07.014

Hu T, and Liao Z. Study on the Integration of Hydrodynamic and Water Quality Model and GIS based on Case Database. Tongji University (2017)

Jiang Y, Cheng B, Liu M, and Nie Y. Spatial and temporal variations of taste and odor compounds in surface water, overlying water and sediment of the Western Lake Chaohu, China. Bull. Environ. Contam. Toxicol. **96**(2), 186–191 (2016). ISSN 0007-4861. https://doi.org/10.1007/s00128-015-1698-y

Jiang Y, Yang Y, Wu Y, Tao J, and Cheng B. Microcystin bioaccumulation in freshwater fish at different trophic levels from the eutrophic Lake Chaohu, China. Bull. Environ. Contam. Toxicol. 1–6 (2017). ISSN 1432-0800. https://doi.org/10.1007/s00128-017-2047-0

Jin H, and Chang Z. The pollution way of microcystins and their bioaccumulation in terrestrial plants: a review. Shengtai Xuebao/Acta Ecol. Sin. **33**(11), 3298–3310 (2013). https://doi.org/10.5846/stxb201203160356

Jin X, Xu Q, and Huang C. Current status and future tendency of lake eutrophication in China. Sci. China. Ser. C Life Sci./Chin. Acad. Sci. **48**(Spec No. 2), 948–954 (2005)

Jørgensen SE. *Ecological Model Types*, vol. 28. Elsevier (2016)

Kang L, He Q-S, He W, Kong X-Z, Liu W-X, Wu W-J, Li Y-L, Lan X-Y, and Xu F-L. Current status and historical variations of DDT-related contaminants in the sediments of Lake Chaohu in China and their influencing factors. Environ. Pollut. **219**, 883–896 (2016a). https://doi.org/10.1016/j.envpol.2016.08.072

Kang L, Wang Q-M, He Q-S, He W, Liu W-X, Kong X-Z, Yang B, Yang C, Jiang Y-J, and Xu F-L. Current status and historical variations of phthalate ester (PAE) contamination in the sediments from a large Chinese lake (Lake Chaohu). Environ. Sci. Pollut. Res. **23**(11), 10393–10405 (2016b). ISSN 0944-1344. https://doi.org/10.1007/s11356-015-5173-4. International Conference on Contaminated Sediments (ContaSed-2015), Ascona, Switzerland, 08–13 Mar 2015

Kong X, He W, Liu W, Yang B, Xu F, Jorgensen SE, and Mooij WM. Changes in food web structure and ecosystem functioning of a large, shallow Chinese lake during the 1950s, 1980s and 2000s. Ecol. Model. **319**(SI), 31–41 (2016). ISSN 0304-3800. https://doi.org/10.1016/j.ecolmodel.2015.06.045

Kong X, He Q, Yang B, He W, Xu F, Janssen ABG, Kuiper JJ, van Gerven LPA, Qin N, Jiang Y, Liu W, Yang C, Bai Z, Zhang M, Kong F, Janse JH, and Mooij WM. Hydrological regulation drives regime shifts: evidence from paleolimnology and ecosystem modeling of a large shallow Chinese lake. Glob. Change Biol. **23**(2), 737–754 (2017). ISSN 1365-2486. https://doi.org/10.1111/gcb.13416

Li T, and Gao X. Ecosystem services valuation of lakeside wetland park beside Chaohu Lake in China. Water **8**(7), 301 (2016). ISSN 2073-4441. https://doi.org/10.3390/w8070301

Liu J, and Yang W. Water sustainability for China and beyond. Science **337**(6095), 649–650 (2012)

Li D, Zuo Q, and Cui G. Disposal of chemical contaminants into groundwater: viewing hidden environmental pollution in China. Environ. Earth Sci. **70**(4), 1933–1935 (2013a). ISSN 1866-6280. https://doi.org/10.1007/s12665-013-2463-3

Li X-G, He H-Y, and Sun Q-F. The shallow groundwater pollution's assessment of west Liaohe plain (eastern). J. Chem. Pharm. Res. **5**(11), 290–295 (2013b)

Li C, Huo S, Yu Z, Guo W, Xi B, He Z, Zeng X, and Wu F. Historical records of polycyclic aromatic hydrocarbon deposition in a shallow eutrophic lake: impacts of sources and sedimentological conditions. J. Environ. Sci. **41**, 261–269 (2016a). ISSN 1001-0742. https://doi.org/10.1016/j.jes.2015.05.007

Li G, Xie F, Zhang J, Wang J, Yang Y, and Sun R. Occurrence of phosphorus, iron, aluminum, silica, and calcium in a eutrophic lake during algae bloom sedimentation. Water Sci. Technol. **74**(6), 1266–1273 (2016b). ISSN 0273-1223. https://doi.org/10.2166/wst.2016.277

Li J, Chen F, Liu Z, Zhao X, Yang K, Lu W, and Cui K. Bottom-up versus top-down effects on ciliate community composition in four eutrophic lakes (China). Eur. J. Protistology **53**, 20–30 (2016c). ISSN 0932-4739. https://doi.org/10.1016/j.ejop.2015.12.007

Li J, Zhang Y, Ma R, Duan H, Loiselle S, Xue K, and Liang Q. Satellite-based estimation of column-integrated algal biomass in nonalgae bloom conditions: a case study of Lake Chaohu, China. IEEE J. Sel. Top. Appl. Earth Observations Remote Sens. **10**(2), 450–462 (2017). ISSN 1939-1404. https://doi.org/10.1109/JSTARS.2016.2601083

Liu K, Li J, Yan S, Zhang W, Li Y, and Han D. A review of status of tetrabromobisphenol A (TBBPA) in China. Chemosphere **148**, 8–20 (2016a). ISSN 0045-6535. https://doi.org/10.1016/j.chemosphere.2016.01.023

Liu C, Shao S, Shen Q, Fan C, Zhang L, and Zhou Q. Effects of riverine suspended particulate matter on the post-dredging increase in internal phosphorus loading across the sediment-water interface. Environ. Pollut. **211**, 165–172 (2016b). ISSN 0269-7491. https://doi.org/10.1016/j.envpol.2015.12.045

Liu W-X, Wang Y, He W, Qin N, Kong X-Z, He Q-S, Yang B, Yang C, Jiang Y-J, Jorgensen SE, and Xu F-L. Aquatic biota as potential biological indicators of the contamination, bioaccumulation and health risks caused by organochlorine pesticides in a large, shallow Chinese lake (Lake Chaohu). Ecol. Indic. **60**, 335–345 (2016c). ISSN 1470-160X. https://doi.org/10.1016/j.ecolind.2015.06.026

Liu Y-L, Wang J-Z, Peng S-C, and Chen T-H. Modeling the environmental behaviors and ecological risks of permethrin in Chaohu Lake. Huanjing Kexue/Environ. Sci. **37**(12), 4644–4650 (2016d). https://doi.org/10.13227/j.hjkx.201606022

Liu C, Zhang L, Fan C, Xu F, Chen K, and Gu X. Temporal occurrence and sources of persistent organic pollutants in suspended particulate matter from the most heavily polluted river mouth of Lake Chaohu, China. Chemosphere **174**, 39–45 (2017). ISSN 0045-6535. https://doi.org/10.1016/j.chemosphere.2017.01.082

Peng L, Liu Y, Chen W, Liu L, Kent M, and Song L. Health risks associated with consumption of microcystin-contaminated fish and shellfish in three Chinese lakes: significance for freshwater aquacultures. Ecotoxicol. Environ. Saf. **73**(7), 1804–1811 (2010). https://doi.org/10.1016/j.ecoenv.2010.07.043

Qian J, Zhao W, Hong T, Lu Y, and Tang C. Spatial variability in hydrochemistry of groundwater and surface water: a case study in Nanfei River catchment, China, in *Proceedings of the 12th International Symposium on Water-Rock Interaction, Kunming, China, 31 July–5 August 2007*, ed. by Y. Wang, T.D. Bullen, vol. 2 (Taylor & Francis, 2007), pp. 887–890

Shang GP, and Shang JC. Spatial and temporal variations of eutrophication in western Chaohu Lake, China. Environ. Monit. Assess. **130**(1–3), 99–109 (2007). ISSN 0167-6369. https://doi.org/10.1007/s10661-006-9381-8

Shang L, Feng M, Liu F, Xu X, Ke F, Chen X, and Li W. The establishment of preliminary safety threshold values for cyanobacteria based on periodic variations in different microcystin congeners in Lake Chaohu, China. Environ. Sci. Processes Impacts **17**(4), 728–739 (2015). ISSN 2050-7887. https://doi.org/10.1039/c5em00002e

Shiguang S, Lianqing X, Cheng L, Jingge S, Zhaode W, Xiang H, and Chengxin F. Assessment of heavy metals in sediment in a heavily polluted urban river in the Chaohu Basin, China. Chin. J. Oceanol. Limnol. **34**(3), 526–538 (2016). ISSN 0254-4059. https://doi.org/10.1007/s00343-015-4240-5

Sun H, Cheng X, and Dai M. Regional flood disaster resilience evaluation based on analytic network process: a case study of the Chaohu Lake Basin, Anhui Province, China. Nat. Hazards **82**(1), 39–58 (2016). ISSN 0921-030X. https://doi.org/10.1007/s11069-016-2178-3

Tang DL, Kawamura H, Oh IS, and Baker J. Satellite evidence of harmful algal blooms and related oceanographic features in the Bohai Sea during autumn 1998. Adv. Space Res. **37**, 681–689 (2006)

Tang W, Shan B, Zhang H, and Mao Z. Heavy metal sources and associated risk in response to agricultural intensification in the estuarine sediments of Chaohu Lake Valley. East China. J. Hazard. Mater. **176**(1–3), 945–951 (2010). https://doi.org/10.1016/j.jhazmat.2009.11.131

Wang SF, Tang DL, and He FL et al. Occurrences of Harmful Algal Blooms (HABs) associated with ocean environments in the South China Sea. Hydrobiologia **596**, 79–93 (2008).

Wang Q, Gu Y, and Sun D. Spatial and seasonal variations of nitrate-N concentration in groundwater within Chao Lake watershed. Shengtai Xuebao/Acta Ecol. Sin. **34**(15), 4372–4379 (2014). https://doi.org/10.5846/stxb201212111779

Wang J-Z, Bai Y-S, Wu Y, Zhang S, Chen T-H, Peng S-C, Xie Y-W, and Zhang X-W. Occurrence, compositional distribution, and toxicity assessment of pyrethroid insecticides in sediments from the fluvial systems of Chaohu Lake, Eastern China. Environ. Sci. Pollut. Res. **23**(11), 10406–10414 (2016a). ISSN 0944-1344. https://doi.org/10.1007/s11356-015-5831-6. International Conference on Contaminated Sediments (ContaSed-2015), Ascona, Switzerland, 08-13 Mar 2015

Wang J-Z, Peng S-C, Chen T-H, and Zhang L. Occurrence, source identification and ecological risk evaluation of metal elements in surface sediment: toward a comprehensive understanding of heavy metal pollution in Chaohu Lake, Eastern China. Environ. Sci. Pollut. Res. **23**(1), 307–314 (2016b). ISSN 0944-1344. https://doi.org/10.1007/s11356-015-5246-4

Weigen J, Chengqin Y, and Seuffert O. Nonpoint pollution controlled by a multi-pond agroecosystem in a subwatershed of Chaohu Lake. China. Geoökodynamik **11**(2–3), 191–212 (1990)

Wohlfart C, Kuenzer C, Chen C, and Liu G. Social-ecological challenges in the Yellow River basin (China): a review. Environ. Earth Sci. **75**, 1066 (2016)

Wu P, Gao C, Chen F, and Yu S. Response of organic carbon burial to trophic level changes in a shallow eutrophic lake in SE China. J. Environ. Sci. **46**, 220–228 (2016). ISSN 1001-0742. https://doi.org/10.1016/j.jes.2016.05.003

Xie J. *Comparative Research of Different Transformation Strategies for Urban Drainage System Based on Model* (Tongji University, College of Environmental Science and Engineering, 2017)

Xie P, and Liu J. Practical success of biomanipulation using filter-feeding Fish to control cyanobacteria blooms: a synthesis of decades of research and application in a subtropical hypereutrophic lake. Sci. World J. [electronic resource] **1**, 337–356 (2001)

Xie L, Xie P, Guo L, Li L, Miyabara Y, and Park H-D. Organ distribution and bioaccumulation of microcystins in freshwater fish at different trophic levels from the eutrophic Lake Chaohu. China. Environ. Toxicol. **20**(3), 293–300 (2005). https://doi.org/10.1002/tox.20120

Xie J, Liao Z, and Gu X. Prediction and evaluation of waterlogging in highly urbanized areas based on Mike Urban: demonstrated on the example of Huoshan-Huimin Drainage System in Shanghai. Energy Environ. Prot. **30**(5), 44–49 (2016)

Xie J, Chen H, Liao Z, Gu X, Zhu D, and Zhang J. An integrated assessment of urban flooding mitigation strategies for robust decision making. Environ. Model. Softw. **95**:143–155 (2017). ISSN 1364-8152. https://doi.org/10.1016/j.envsoft.2017.06.027

Xu F-L, Yang C, He W, He Q-S, Li Y-L, Kang L, Liu W-X, Xiong Y-Q, and Xing B. Bias and association of sediment organic matter source apportionment indicators: a case study in a eutrophic Lake Chaohu, China. Sci. Total Environ. **581–582**, 874–884 (2017). ISSN 0048-9697. https://doi.org/10.1016/j.scitotenv.2017.01.037

Xue K, Zhang Y, Duan H, and Ma R. Variability of light absorption properties in optically complex inland waters of Lake Chaohu, China. J. Great Lakes Res. **43**(1), 17–31 (2017). ISSN 0380-1330. https://doi.org/10.1016/j.jglr.2016.10.006

Yang H, Tang J, Hu A, Yang Y, and Xie L. Identification and denitrification characteristics of a denitrifier. Chin. J. Environ. Eng. **8**(1), 366–371 (2014)

Yang B, Jiang Y-J, He W, Liu W-X, Kong X-Z, Jorgensen SE, and Xu F-L. The tempo-spatial variations of phytoplankton diversities and their correlation with trophic state levels in a large eutrophic Chinese lake. Ecol. Indic. **66**, 153–162 (2016a). ISSN 1470-160X. https://doi.org/10.1016/j.ecolind.2016.01.013

Yang F, Geng D, Wei C, Ji H, and Xu H. Distribution of arsenic between the particulate and aqueous phases in surface water from three freshwater lakes in China. Environ. Sci. Pollut. Res. **23**(8), 7452–7461 (2016b). ISSN 0944-1344. https://doi.org/10.1007/s11356-015-5998-x

Yang Z, Kong F, and Zhang M. Groundwater contamination by microcystin from toxic cyanobacteria blooms in Lake Chaohu, China. Environ. Monit. Assess. **188**(5), 280 (2016c). ISSN 0167-6369. https://doi.org/10.1007/s10661-016-5289-0

You D-J, Chen X-G, Xiang H-Y, Ouyang L, and Yang B. Isolation, identification and characterization of a microcystin-degrading bacterium Paucibacter sp. strain CH. Huanjing Kexue/Environ. Sci. **35**(1), 313–318 (2014)

Yu L, Kong F, Zhang M, Yang Z, Shi X, and Du M. The dynamics of Microcystis genotypes and microcystin production and associations with environmental factors during blooms in Lake Chaohu, China. Toxins **6**(12), 3238–3257 (2014). ISSN 2072-6651. https://doi.org/10.3390/toxins6123238

Yue T-X, Xu B, and Zhao N et al. Thematic issue: environment and health in China-I. Environ. Earth Sci. **74**(8), 6361–6365 (2015)

Zan F, Huo S, Xi B, Li Q, Liao H, and Zhang J. Phosphorus distribution in the sediments of a shallow eutrophic lake, Lake Chaohu, China. Environ. Earth Sci. **62**(8), 1643–1653 (2011). ISSN 1866-6299. https://doi.org/10.1007/s12665-010-0649-5

Zan F, Huo S, Xi B, Zhu C, Liao H, Zhang J, and Yeager KM. A 100-year sedimentary record of natural and anthropogenic impacts on a shallow eutrophic lake, Lake Chaohu. China. J. Environ. Monit. **14**(3), 804–816 (2012)

Zhang G, and Liao Z. Research and Application of Hydrodynamic and Water Quality Model in Small and Medium Cities of Inland, a case study of Chaohu City (2016)

Zhang Z, and Gao J. Linking landscape structures and ecosystem service value using multivariate regression analysis: a case study of the Chaohu Lake Basin. China. Environ. Earth Sci. **75**(1), 3 (2016)

Zhang L, Bai Y-S, Wang J-Z, Peng S-C, Chen T-H, and Yin D-Q. Identification and determination of the contribution of iron–steel manufacturing industry to sediment-associated polycyclic aromatic hydrocarbons (PAHs) in a large shallow lake of eastern China. Environ. Sci. Pollut. Res. **23**(21), 22037–22046 (2016a). ISSN 1614-7499. https://doi.org/10.1007/s11356-016-7328-3

Zhang L, Yin D-Q, Wu Y, Peng S-C, Chen T-H, and Wang J-Z. Organochlorine pesticides in sediments around Chaohu Lake: concentration levels and vertical distribution. Soil Sediment Contam. **25**(2), 195–209 (2016b). ISSN 1532-0383. https://doi.org/10.1080/15320383.2016.1112767

Zhang M, Zhang Y, Yang Z, Wei L, Yang W, Chen C, and Kong F. Spatial and seasonal shifts in bloom-forming cyanobacteria in Lake Chaohu: patterns and driving factors. Phycol. Res. **64**(1), 44–55 (2016c). ISSN 1440-1835. https://doi.org/10.1111/pre.12112

Zhi G, Chen Y, Liao Z, Walther M, and Yuan X. Comprehensive assessment of eutrophication status based on Monte Carlo–triangular fuzzy numbers model: site study of Dongting Lake, Mid-South China. Environ. Earth Sci. **75**(12), 1011 (2016). ISSN 1866-6299. https://doi.org/10.1007/s12665-016-5819-7

Zhou Y, and Liao Z. *Review of GIS-Based Domestic Urban Water Environment Management* (Environmental Science and Management, Sept, 2014)

Zhou Q, Liu C, and Fan C. Application of plow-tillage as an innovative technique for eliminating overwintering cyanobacteria in eutrophic lake sediments. Environ. Pollut. **219**, 425–431 (2016). ISSN 0269-7491. https://doi.org/10.1016/j.envpol.2016.05.026

Chapter 2
Managing Water Resources for Urban Catchments

Olaf Kolditz, Thomas U. Berendonk, Cui Chen, Lothar Fuchs, Matthias Haase, Dirk Jungmann, Thomas Kalbacher, Peter Krebs, Christian Moldaenke, Roland Müller, Frank Neubert, Karsten Rink, Karsten Rinke, Agnes Sachse and Marc Walther

2.1 Objective

The overall objective of the project was the development of water management system solutions for a sustainable improvement of water quality in the city of Chaohu and in the Chao Lake. The Urban Water Resources Management (UWRM) concept is the innovative approach, which includes both efficient urban water management in urban and suburban areas, as well as interaction with aquatic ecosystems. Data and models for planning purposes and regional water management are made available by using a comprehensive online environmental information system for authorities and water suppliers. The Chao Lake plays a central role as an ecological and economic

---

O. Kolditz (✉) · C. Chen · K. Rink · M. Walther
Department of Environmental Informatics, Helmholtz Centre of Environmental Research–UFZ, Permoserstraße 15, 04318 Leipzig, Germany
e-mail: olaf.kolditz@ufz.de

C. Chen
e-mail: cui.chen@ufz.de

K. Rink
e-mail: karsten.rink@ufz.de

M. Walther
e-mail: marc.walther@ufz.de

O. Kolditz
Applied Environmental System Analysis, Technische Universität Dresden, Dresden, Germany

T. U. Berendonk · D. Jungmann
Department of Hydrosciences, Institute of Hydrobiology, Chair of Limnology, Technische Universität Dresden, Zellescher Weg 40, 01217 Dresden, Germany
e-mail: thomas.berendonk@tu-dresden.de

L. Fuchs
Institute for Technical and Scientific Hydrology, Engelbosteler Damm 22, 30167 Hannover, Germany
e-mail: L.Fuchs@itwh.de

© Springer Nature Switzerland AG 2019
A. Sachse et al. (eds.), *Chinese Water Systems*, Terrestrial Environmental Sciences, https://doi.org/10.1007/978-3-319-97568-9_2

protection and raw water supplier for the drinking water supply of the population of the city of Chaohu. The research and development project (R&D Project) thus makes an important contribution to the sustainable development of the Chaohu region as part of the Masterplan Ecological Seascape Chaohu of the Anhui Provincial Government. The scientific and technical solutions are implemented in demonstration projects.

2.2 Proposed Solution

In order to achieve the project objective, methods of the UWRM concept were applied, such as the holistic view of the urban water network and all levels of the aquatic system according to the principle of emissions (pollution source) / immissions (contamination). Urban water management includes a comprehensive view of all urban resources: the lake as a source of drinking water and objects worthy of protection, the restoration of the urban water network, rainwater management and wastewater treatment. Rapid development rates in the demonstration region fully take into account suburban and rural areas and the impact of agriculture (non-point sources of pollution). The UWRM concept is based on a regional implementation strategy with flexible decentralized cluster solutions for wastewater treatment. An important basis for the successful implementation of the UWRM concept is the establishment of extensive monitoring platforms (Cities and Lakes Observatories) for sources of water pollution and the recipient Chao Lake. The observatories serve as an early warning system for operational water management (drinking water).

M. Haase
WISUTEC Umwelttechnik GmbH, Jagdschänkenstraße 50, 09117 Chemnitz, Germany
e-mail: m.haase@wisutec.de

T. Kalbacher · A. Sachse
OpenGeoSys e.V, Lampestraße 5, 04107 Leipzig, Germany
e-mail: agnes.sachse@web.de

P. Krebs
Department of Hydrosciences, Institute of Urban and Industrial Water Management, Technische Universität Dresden, Bergstraße 66, 01069 Dresden, Germany
e-mail: peter.krebs@tu-dresden.de

C. Moldaenke
bbe Moldaenke GmbH, Preetzer Chausee 177, 24222 Schwentinental, Germany
e-mail: cmoldaenke@bbe-moldaenke.de

R. Müller
Centre for Environmental Biotechnology, Decentralized Wastewater Treatment and Reuse, Helmholtz Centre of Environmental Research–UFZ, Permoserstraße 15, 04318 Leipzig, Germany
e-mail: roland.mueller@ufz.de

F. Neubert
AMC–Analytik & Messtechnik GmbH Chemnitz, Heinrich-Lorenz-Straße 55, 09120 Chemnitz, Germany
e-mail: Frank.Neubert@amc-systeme.de

Long-term monitoring also allows reliable evaluation of measures and observations of pollutants such as micropollutants. An Environmental Information System (EIS) provides data for the UWRM concept, including required data infrastructures, interoperable simulation tools and web services. On the one hand, the combination of monitoring and modelling platforms in the EIS enables the identification of pollutant sources and pathways in the entire catchment area and, on the other hand, is an important instrument for operational water management and long-term water quality forecasts. The modelling platforms examine all levels of the coupled hydrological system including soils and groundwater.

The implementation concept was developed in close cooperation with regional stakeholders and municipal utilities. The demonstration projects - in which four German companies are involved - play an important role in the cooperation between industry and science and the actual applicability of the results. The project objectives, the methodology and the implementation concept have resulted in the following project structure for the R&D project "Urban Catchments", which are divided into 6 subprojects:

- A. Urban water management
- B. Decentralized wastewater management
- C. Chao Lake
- D. Environmental information system
- E. Groundwater
- Z. Project Management

In response to programs already initiated in the Chaohu community, representatives of the Tongji Research Institute recommended that the model region focus on the city of Chaohu and the east side of the lake for the R&D Project "Urban Catchments". Due to rapid urbanization in the region, suburban and rural areas are included for scalable decentralized sewerage techniques and for shutting down the water/mass balances of Chao Lake to model the entire catchment area.

The project structure is regionally divided into the urban (A) and suburban/rural areas (B) as well as Chao Lake (C) with the environmental information system acting as an integrating element of the project (D). Figure 2.1 is a schematic diagram of the project structure, which includes methodical elements as well as regional elements like the central asset Chao Lake and the observatories. Efficient management of water resources required a monitoring network and modelling tools to plan and create appropriate measures.

K. Rinke
Department of Lake Research, Helmholtz Centre of Environmental Research–UFZ, Brückstraße 3a, 39114 Magdeburg, Germany
e-mail: karsten.rink@ufz.de

M. Walther
Department of Hydrosciences, Institute for Groundwater Management, Professorship of Contaminant Hydrology, Technische Universität Dresden, Bergstraße 66, 01069 Dresden, Germany

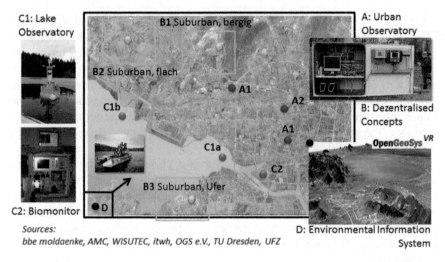

Fig. 2.1 Work packages overview, http://www.ufz.de/urbancatchments/index.php?en=34132

In addition to technological and solution-oriented subprojects of the larger R&D project, the discussion on the cross-cutting issue of capacity development (CD) was continued. The most important aspects in CD can be found in all sub-projects in the areas of plant design, operation and maintenance; Quality control; Data management and software training in the use of the environmental information system (monitoring and modeling).

2.3 Background

Rapid economic development and population growth in China go hand-in-hand with increasing urbanization, involving growing mega-cities, industrialization, and intensified agriculture. As a result, soils and water resources are increasingly stressed and productive management strategies towards sustainable planning are urgently needed. Pollution containing hazardous substances for environmental and human health, depletion of water resources as a result of demographic change, and urbanization processes are increasingly at an alarming rate. Consequently, the protection of aquatic ecosystems and the security of drinking water provision are becoming an increasing economic challenge in water management. The Chinese government recognizes the importance and complexity of the situation and has initiated a program titled Major Water Program of Science and Technology for Water Pollution and Governance (2006–2020) (see Sects. 2.3.1 and 2.3.3). Chao Lake currently exhibits extremely high levels of pollution (some areas are at the worst Level V). The causes are not only the lacking capacity of water treatment used in industry and intense regional agriculture, but also the lack of integrated plans for resilient environmental engineering for urban and rural areas. Many scientific studies have examined the most prevalent pollutants–lake and river sediments contain sometimes high levels of heavy metals, alkylbenzene, and pesticides–but research on technical solutions to improve the water quality has not been substantial. Summer algal blooms in Chao Lake are regular occurrences due to extremely high phosphorus imports, which then results in high levels of cyanotoxins (microcystin, in particular). Chao Lake is the most important source of drinking water in the area, and this severe human health hazard presents an urgent need for action towards improving the water quality. The BMBF-project "Managing Water Resources for Urban Catchments in Chaohu city and Chao Lake" presented here set itself the goal of developing versatile solution strategies using monitoring and model concepts.

Valuable insights and foundations for the development of the Chao Lake region as well as the need for research and measures for solutions to improve the water quality of the Chao Lake have already been provided by the BMBF joint project "Minimization of eutrophication at Lake Chao, China - Anhui Province" (2005–2010). Partners included TU Braunschweig (Leichtweiß-Institut für Hydraulic Engineering, Institute of Sanitary and Environmental Engineering), the Leibnitz-Institute of Freshwater Ecology and Inland Fisheries (IGB Berlin), the University of Jena (Vorhaben 02WT0529) and the Anhui Environmental Protection Bureau (AEPB). This project focused primarily on Heifei drinking water treatment (biologically activated filtration to remove cyanotoxins prior to actual drinking water treatment), lake eutrophication (including macrophyte purification techniques, such as reed stocks, as a "green liver") and the development of a decision support tool (Krüger et al. 2010, 2012). There is a key gap in dealing with urban areas in terms of urban water resource management (UWRM). In many countries, like China, there is a large division between sanitary engineering (water provision and sanitary sewers) and river and lake management. The role of the urban environment plays on the ecological status, the functioning of and the services provided by the aquatic ecosystems is not being adequately considered. This gap should be filled by integrating engineering expertise into the "Urban Catchments"-project.

2.3.1 "Mega Water Program"

The "Mega Water Program" cooperation is a joint declaration of intent for German-Chinese cooperation. The "Major Program of Science and Technology for Water Pollution Control and Governance" ("Mega Water Program") brings together a number of different Chinese ministries with the goal to reach a significant improvement of water quality in selected lakes and rivers by 2020. The Innovation Cluster "Mega Water Program"[1] is a project supported by the Federal Ministry of Education and

---

[1] http://sino-german-major-water.net/de/.

Research (BMBF) in the program "CLIENT - International Partnerships for Sustainable Climate Protection and Environmental Technologies and Services" in the field of "Sustainable Water Management" since 2015. The Innovation Cluster "Mega Water" includes the R&D project "Clean Water from Source to Consumer" (SIGN), "Good Water Governance and German Water Technology for Two Important Chinese Waters"(SINOWATER) and "Managing Water Resources for Urban Catchments: Pilot Project Chao Lake" (URBAN CATCHMENTS). The aim of the Innovation Cluster "Mega Water Program" is to intensify Sino-German research and economic cooperation and to improve the opportunities for the German companies involved in the projects in the Chinese market. The overall network, with the speaker Prof. Dr.-Ing. Max Dohmann of the BMBF, includes the main topics:

- Urban Water Management,
- Integral urban drainage and rainwater treatment,
- Wastewater treatment and sewage sludge disposal,
- Monitoring programs, early warning concepts and environmental information systems,
- Analysis of water eutrophication, sea and river remediation,
- Evaluation of pollutant degradation and ecotoxicology,
- Development of water treatment processes,
- Operation of water distribution systems as well
- "Good Water Governance".

The coordination of the Innovation Cluster Mega Water in China is supported by the BMBF-project office "Clean Water" (see Sect. 2.3.2) in Shanghai.

2.3.2 *"Clean Water"*

Since July 2012 the project office "Clean Water" at Tongji University in Shanghai has been supporting the implementation of the Sino-German "Research and Innovation Programme Clean Water" as well as bilateral water research projects in China, which are implemented specifically through the CLIENT Fund.[2] The project office supports, in co-operation with its Chinese partners, relevant decision-makers in identifying problems and needs in the Chinese water sector as well as the finding and promotion of innovative solutions and their implementation. Furthermore the project office supports the formation of research co-operation between universities and the private sector and facilitates access to Chinese institutions.

---

[2]https://www.fona.de/en/project-office-clean-water-16846.html.

2.3.3 13th Five-Year-Plan of Major Water Project in Chaohu Basin

The Chinese Government released its 13th Five-Year Plan (2016–2020)[3] on 17th March 2016. It promotes a cleaner and greener economy, with strong commitments to environmental management and protection, clean energy and emission control, ecological protection and security, and the development of green industries. Specific objectives for environmental protection in the 13th Five Year Plan (FYP) period include:

- reduction of water consumption by 35% by 2020 as compared to 2013;
- estimated total consumption of primary energy in 2020 of less than 5 billion tons of standard coal;
- energy consumption per unit of GDP to be reduced by 15% in 2020 (compared to 2015);
- reduction of carbon dioxide emissions per unit of GDP by 40–45% by 2020 (compared to 2015 which is consistent with China's Plan for Addressing Climate Change (2014–2020)).

The 13th FYP of Major Water Project on Chaohu Basin aims firstly to establish an index system and methods for evaluating and selecting the pollution control technologies applicable to Chaohu Basin. The plan will innovate an integrated governance mode for the entire process of the small watershed pollution control which can reduce the pollutant load of Chao Lake by the clean runoff and water conservation in the upper reaches of Paihe River. The 13th FYP is, with regard to the Chaohu Basin, divided into five sub-projects:

1. Subproject 1 focuses on the application and verification of the most important watershed pollution control techniques in the Chaohu Basin. The aim is to achieve the objectives of good water quality for the Chao Lake and to develop a scheme that optimizes the reduction of pollutants in the catchment area. In addition, a water quality target management and decision-making platform for the entire Chaohu Basin will be developed to formulate a comprehensive and sustainable use of Lake Chao Lake for the city of Hefei City.
2. Subproject 2 includes the integration of water conservation and environmental protection technologies for clean water catchment areas. It is about the Jianghuai area, where agriculture is a diffuse source of major pollution in the upstream watershed of the Paihe River. Techniques need to be developed to control the discharge of nitrogen and phosphorus both in the water cycle and in agriculture and to prevent the transfer of pollutants to the nearby wetlands. These techniques must be integrated into the upper reaches of the Paihe River. The result of this subproject would be an innovative model for clean drainage and water conservation in agriculture and the control of urban pollution and outflow regulation.

---

[3]http://www.gov.cn/xinwen/2016-03/17/content5054992.htm.

3. Subproject 3 focuses on small catchment areas in the Chaohu Basin with diversification of pollution types, diversity of land use, interdependence of urban and rural areas, verification and improvement of urban and suburban water treatment techniques, groundwater abstraction technologies as well as river-ecological rehabilitation, and urban non-point pollution.
4. Subproject 4 focuses on ensuring water quality and reducing pollutants from inputs like agriculture and the inflow of River Paihe, especially in the drinking water corridors. It will also promote technologies that improve water quality.
5. Subproject 5 aims to review and improve the Chao Lake water body, and develop long-term treatment programs that prevent algal blooms using biological and physical methods. In addition, long-term technical measures for eutrophication regulation are being prepared in Chao Lake.

The "Urban Catchments"-project is based on the approaches of both the 11th and 12th as well as the 13th Five Year Plan, thereby supporting the model- and technology-based realization of environmental objectives in the Chaohu Basin.

2.4 "Urban Catchments"-Project History

The successful cooperation in the "Urban Catchments"-project could only be realised through the ongoing project meetings, workshops and research trips. The following is a brief outline of the events and was documented by the UC Project Management on the Project Management Wiki: https://svn.ufz.de/urbancatchments. In the following, the most important events are presented chronologically.

2.4.1 Important Dates and Achievements in 2015

On 01.04.2015, the BMBF-CLIENT project "Managing Water Resources for Urban Catchments" (www.ufz.de/urbancatchments) was officially launched. Immediately afterwards, the first network meeting took place in Bonn. Ministerialrat Dr. Ulrich Katenkamp invited the three German-Chinese CLIENT joint projects SIGN, SINOWATER and URBAN CATCHMENTS to the first coordination and networking meeting on 13 April 2015 in the Federal Ministry of Education and Research in Bonn. This meeting of all German project partners served to coordinate and network with each other. At the same time, preparations were made for the Chinese kick-off meeting in Beijing. A few days later, the first project meeting took place in Leipzig (28.04.2015).

The project Kick-off in Beijing took place on 06–07.05.2015 (Fig. 2.2): State Secretary Dr. Georg Schütte and Vice Minister Prof. Jianlin Cao sign a joint declaration on the cooperation "Mega Wasser Programm" to kick off the work of the three CLIENT joint projects funded by the BMBF. In order to exploit technical synergies

Fig. 2.2 Joint-Declaration and Kick-Off in Beijing (Source: BMBF/MoST)

in support of the impact of German research and development projects in China, the three CLIENT collaborative projects have been combined into one Major Water Innovation Cluster. As German spokesman of the Major Water Innovation Cluster, Ministerialrat Dr. Ing. Ulrich Katenkamp introduced Prof. Dr.-Ing. Max Dohmann. On the Chinese side, Prof. Yonghui Song replaced Prof. Meng Wei, President of the Chinese Research Academy for Environment and Science (CRAES). The three German-Chinese CLIENT collaborative projects SIGN, SINOWATER and URBAN CATCHMENTS met the day before to get to know each other across the borders of the individual projects. Networking topics were discussed in a "World Café" with the Chinese project partners under the headings: Good Water Governance + Lake Eutrophication, Lake River Conservation + Monitoring, Modelling, Early Warning Concepts, Environmental Information Systems, + Urban Water Management, Sewerage, Stormwater, wastewater treatment, sludge disposal + water treatment/water distribution.

On 29.06.2015 the 2nd Project Meeting took place in Leipzig with discussions on cooperation agreements and working plans of the working packages (Fig. 2.3).

On July 10, 2015, a meeting of industrial partners within the "Urban Catchments"-project like WISUTEC, AMC, bbe, TUD-HYB took place in Chemnitz. The aim of the meeting was the detailed coordination of the activities of all subproject partners until the summer of 2016: times agreed for the achievement of milestones. The first task is a functioning pilot data network in Germany to demonstrate on the one hand the connectivity of the devices and probes and tests and on the other hand to show the possibilities of software and hardware solutions to the Chinese delegation from Chaohu. In addition, requirements were discussed at Chinese side

Fig. 2.3 2nd Project Meeting in Leipzig (Source: UC Project)

and will be regulated by contract: starting work on site, site infrastructure at the site (electricity, water, network, work, laboratory material, etc.), access to the site (for project staff and service providers), maintenance (frequency, activity) during ongoing operation, certification of software and hardware (including reimbursement of costs), payment of additional costs, e.g. Mobile costs. The data transmission on site was again discussed. Based on the discussions of the last project meeting, it was determined that the buoy regularly loads its data onto a still-to-be-defined FTP server (presumably with Chao Lake Management Authority (CLMA)). AMC will operate (presumably virtualised) a server with a SensoMASTER service at the CLMA, which will retrieve the data from the buoy and forward it to the data center. In the drinking water works, the computer of the daphnia toximeter is equipped with the SensoMaster software as a data collector and acts locally as a data collector and transmitter. Due to the high hardware requirements of the daphnia oximeter (real-time image processing), this solution must first be tested in practice in Germany.

A short time later a meeting took place together with WISUTEC, AMC, UFZ in Chemnitz. Presentation of the water information system for Saxony by WISUTEC and the next activities of working package D: Data integration: GIS data prepared by UFZ ENVINF, integration of GIS data from itwh into the OGS DataExplorer, defining data (IO: parameters, results) for the lake models.

On July 22, 2015, the UFZ, WISUTEC, AMC met in Leipzig. The UFZ data management concept including data logger portal, mass data portal and research portal was presented by WKDV. In addition, the setting up of the FTP server for the UFZ buoy was discussed.

For one week (28.09–03.10.2015) representatives of the Chao Lake Management Authority and Chinese Research Partners met in Leipzig for the BMBF CLIENT "Urban Catchments"-project (Fig. 2.4a). The meeting starts on 28.09.2015 with talks which are given by Prof. Olaf Kolditz, Mr. Tang Xiaoxian, Prof. Zhaosheng Chu, Mrs. Ursula Schmitz, Prof. Chengxin Fan, Prof. Zhenliang Liao, Prof. Peter

Fig. 2.4 (a) Participants of the 1st Chao Lake Workshop in Leipzig (28.09.2015). (b) Meeting of workshop participants in front of the building of Wisutec in Chemnitz (Source: UC Project)

Krebs, Mr. Dr. Manfred van Afferden and Mr. Dr. Frank Neubert. On the second day, a delegation of representatives of the Chao Lake Management Authority visited the project partner WISUTEC Umwelttechnik GmbH in Chemnitz (Fig. 2.4b). They were accompanied by Chinese scientists involved in the "Urban Catchments"-project. The two Chemnitz-based companies WISUTEC Umwelttechnik GmbH and AMC presented modern data management technologies for the water industry. Using the example of the software solutions AL.VIS and SensoMaster of the two SMEs operating in Germany, it was demonstrated how future integrated data management could be designed for Chao Lake. In addition, a meeting took place at the Technische Universität Dresden to discuss the potential supports of TUD needed for online and biomonitoring installation in Chaohu and for monitoring station maintenance and the visit of the online and biomonitoring stations in Dresden. The following day, the UFZ field site in Harz Mountains (Rappbode Reservoir) was visited. The next day, the itwh in Hannover was on the plan with introduction of itwh and related projects and simulation models, talks about data request (surface data, sewer network data, special structures, rainfall data, …), data preparation and explanation of the German regulations and standards. At the end of the week was still a visit to bbe in Kiel (bbe Moldaenke) on the program with presentation of bbe's contribution to the UC project: Early warning for taste and odour problems, Biomonitoring, Chlorophyll probes, new approaches for cyanobacteria measurements and photosynthetic activity, UV fluorescence: monitoring of organic materials determination and Discussion, evaluation of particular Chinese needs and potential application of bbe's technologies close to Chao Lake.

The signing ceremony of the cooperation agreement between UFZ and CLMA took place during the visit of the German Chancellor Angela Merkel city of Hefei in China (26–30.10.2015) (Fig. 2.5). This was an important milestone for the Sino-German cooperation within the Major-Water Program. Germany and China, represented by the Chinese Prime Minister Li Keqiang, have agreed to work together more closely in urban water management. The Chao Lake, which supplies the fast growing cities of Hefei and Chao with drinking water, is particularly heavily burdened by sewage. A "biomonitoring" as an innovative observation method, developed by the Helmholtz Centre for Environmental Research GmbH-UFZ in Leipzig, is to be used

Fig. 2.5 The signing ceremony of the cooperation agreement between UFZ and CLMA (Source: Bundesregierung/Jesco Denzel)

for the first time in this region and to contribute to the improvement of water quality in the long term. The continuously collected measurement data should be fed into an environmental information system. This allows predictions for the overall system "Lake City" and thus an early warning system for the water supply from Chao Lake. This collaboration has now been officially decided. The new cooperation builds a bridge between the participating ministries of research, science and companies, and regional administrations. "In order to bring German know-how from science and industry into the development and implementation of innovations to solve Chinese water problems, we need to involve all stakeholders. This is how both sides benefit", said Federal Research Minister Prof. Dr. Johanna Wanka in Berlin and emphasized that this was also an important step in the context of the new China strategy of the BMBF. The Federal Ministry of Education and Research presented the China strategy to the public on Wednesday. The "Urban Catchments"-project of the Leipzig Helmholtz Centre is part of the cluster of three projects "Mega Water", which is connected to the Mega-Water Program of the Chinese government.

At the same time, the 4th Chaohu workshop took place (28–30.10.2015) in Nanjing, Chaohu (Fig. 2.6), Hefei and Shanghai. The aim of the workshop was to visit the suburban area of Chaohu, discussion and presentation of results within every work groups and excursion to water supply.

Fig. 2.6 4th Chaohu Workshop (Source: Manfred van Afferden)

On 09.12.2015 the UFZ, WISUTEC and representatives of the TU Dresden met in Dresden to discuss the development of the prototype for the environmental information system "Urban Catchments" and to organise the planned presentation in the summer for the next workshop in Chaohu and other potential buyers in Shanghai and Wuhan.

2.4.2 Important Dates and Achievements in 2016

At the end of 2015 respectively at the beginning of 2016, all members of the working groups met each other. The work group A gave a summary of the retrospective 2015 and presented the plans for 2016. For the planned establishment of the Urban Monitoring System in Chaohu, there are delays in identifying the exact points of contact for the selection of suitable locations. In order to avoid delays in project processing, alternatives are therefore found at other locations (Shanghai, Nanjing, Wuhan). The work group B discussed the provision of the FTP server for the UFZ buoy as a prerequisite for the remote sensing data transfer: buoy-> UFZ-> WISUTEC. This is especially important for the UC demonstration project "Environmental Information System". For the demonstration project, the functionality of the UIS prototype for the three data sources has to be presented: (a) buoy in the Rappbodetalsperre (UFZ), (b) Urban Observatory in Dresden (Lockwitzbach) (TUD-SWW) and (c) Biomonitor in Dresden (TUD-HYB), Moldaenke) on the basis of German locations. The next steps of the WP-D were discussed: The work group should focus on the fact that the remote data transmission from the monitoring project "buoy" and the biomonitor to WISUTEC is realized and thus the online data in AL.VIS. This would enable all UC monitoring projects at the German sites to be mapped in the WISUTEC software and important milestone of WP-D in data integration would have been achieved. Since a direct transfer of environmental data abroad is not possible, a corresponding FTP server at NIGLAS should be set up to ensure the data transfer from the buoy to the server. From there, the data can then be obtained for the German partners. Currently the buoy is still in the Rappbodetalsperre. The data transfer from the buoy via the FTP server to the UFZ Data Management Portal (DMP) is operational (as of 01/2016). For the Urban Observatory, the first two (out of 5) stations will first be set up on the grounds of the Tongji University campus. The technical data transmission from the stations to WISUTEC is already working. After consultation with bbe Moldaenke, the UC biomonitor is to be installed in the Wuxi waterworks. For the technical data transmission from the biomonitor to WISUTEC, a software update from AMC has to be done on the biomonitor.

The delegation from Ministry of Science and Technology of China (MOST) led by Mr. Zhe Yang (Deputy General Director for the Major Programs) and Prof. Xiaohu Dai (Dean of the Faculty of Environmental Research at Tongji University) visited UFZ on 12th and 13th January 2016. During this visit, the joint cooperation agreement between UFZ and Tongji University has been renewed under the new challenges in particular with regard to intensive cooperation under the Chinese Government's 13th Five-Year Plan (2016–2020) (Fig. 2.7). The joint research agenda is to continue the development of "Urban Catchments" concepts for integrated water management of lake cities in the North-China plain such as Chaohu.

The delegation talked about the research portfolio Helmholtz Centre for Environmental Research - UFZ (Figs. 2.8 and 2.9). Prof. Georg Teutsch (scientific director of the UFZ) presented the diverse ongoing activities of the German-Chinese cooperation at the "Centre for Advanced Water Research" (CAWR) of the UFZ and the TU Dresden and discussed the prospects of a strategic cooperation with China in environmental research.

On March 15th, 2016 a meeting of the project management of "Urban Catchments" with colleagues of the Institute of Hydrobiology of the Chinese Academy

Fig. 2.7 The cooperation agreement between UFZ and Tongji University (Prof. Dai) has been renewed (Source: UFZ)

Fig. 2.8 The delegation from the Chinese Ministry of Science and Technology (MoST) visited the UFZ in Leipzig and held intensive talks on the prospects of strategic cooperation in environmental research (Source: UFZ)

Fig. 2.9 Subsequently, the Chinese delegation looked at a demonstration of the Chao Lake Environmental Information System (EIS) at the Visualization Centre of the UFZ - VISLAB (Source: UFZ)

of Sciences (CAS-HYB) in Wuhan took place. During a workshop, Prof. Kolditz presented the BMBF-CLIENT project and demonstrated the results of the "Chaohu Environmental Information System" to the professors and students with the aid of a mobile visualization unit. CAS-HYB is project partner of "Urban Catchments" and cooperates in particular with the TU Dresden (Institute of Hydrobiology, Prof. Berendonk). Common interest is especially the biomonitoring. At the end of May 2016, colleagues from TU Dresden (D. Jungmann and M. Rybicki) and the company bbe Moldaenke will start a "Biomonitoring tour" at various facilities in Wuhan, Nanjing and Shanghai (CAS-HYB, Tongji University, NIGLAS) to promote the new methodology/technology and strategies for early warning systems for the continuous monitoring of water quality, for example in the field of drinking water monitoring, in China.

On March 14, 2016, representatives from Chaohu City, Tongji University, Chinese companies, and "Urban Catchments" met for talks in Chaohu on the planned establishment of an "Environmental Data Centre Chaohu". After being welcomed by Mr. Zhang, the head of the Chaohu Construction Bureau, Prof. Liao and his colleagues presented the results of the Chaohu project at Tongji University as part of the Major Water Program (Urban Drainage System and Water Quality Modeling). Subsequently, Prof. Kolditz reported on the BMBF-CLIENT project "Managing Water Resources for Urban Catchments - Chaohu" and demonstrated the current state of data integration with the aid of a mobile visualization device (Fig. 2.10). The discussion was chaired by Mr. Wang (Chaohu Construction Bureau). In particular, the visit served to inform local users and stakeholders about the concepts and opportunities for environmental information systems developed through urban catchments on one side and on the other side to identify opportunities for cooperation between German and Chinese companies in the area of environmental information technology.

Fig. 2.10 O. Kolditz demonstrates the environmental information system Chaohu (part of the BMBF-CLIENT project "Urban Catchments") with a mobile 3D visualization device (Source: N. Umlauf)

Fig. 2.11 The first status seminar of the BMBF-CLIENT project "Managing Water Resources for Urban Catchments" took place in Leipzig (Source: UC Project)

Therefore, the portfolios of the German companies involved in "Urban Catchments" (AMC, bbe Moldaenke, itwh, WISUTEC), technologies and products, equipment and software solutions were presented and contact details were exchanged.

Almost exactly one year after the project was launched, the first status seminar of the BMBF-CLIENT project "Managing Water Resources for Urban Catchments" took place in Leipzig on 19.04.2016 (Fig. 2.11). In the first year of the project, already presentable results were achieved: first models of urban water system (WP-A) and lake (WP-C), progress in the development of planning tools (WP-B) and biomonitoring (WP-C) and a first version of the Environmental Information System (WP-D) with online data links to the UC monitoring systems and 3D visualization. The cooperation with the Chinese partners (CLMA in Chaohu, Tongji University in Shanghai, NIGLAS Nanjing, CAS Hydrobiology in Wuhan) has reached a new level through intensive working journeys - with first joint publications and concrete plans for joint ventures. An important milestone was the "Germany Tour" of the Chinese cooperation partners to all locations of the "Urban Catchments" team (from Dresden to Kiel).

In mid-May 2016, a delegation from the "Urban Catchments"-project, consisting of Dr.-Ing. Dirk Jungmann and Dr. Marcus Rybicki (TU Dresden), Dr. Cui Chen and Thomas Aubron (UFZ Leipzig) as well as Christian Moldaenke from bbe Moldaenke visited various scientific institutes in China (Fig. 2.12a–c). The aim of the trip was to deepen the cooperation with the Chinese institutes and to stimulate research opportunities within the framework of German-Chinese cooperation. The first station led the delegation to the Institute of Hydrobiology of the Chinese Academy of Sciences in Wuhan. A field station was visited at Biandantang Lake near Wuhan and a joint workshop was held. The topic was specifically the application of dynamic (technical) biomonitoring as well as the integration of other online monitoring techniques into the research of the institute. As a result, it was decided to establish a water quality monitoring in a body of water monitored by the institute. The second station was the Nanjing Institute of Geology and Limnology of the Chinese Academy of Sciences (NIGLAS). In addition to the existing cooperation in the area of maritime modelling, the workshop resulted in possible cooperation in the area of 1) sediment

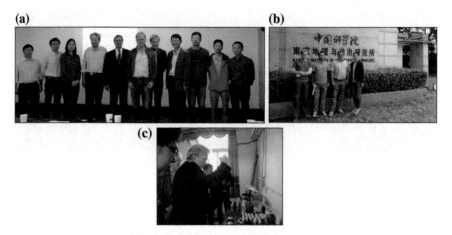

Fig. 2.12 (a) "Urban Catchments" Delegation in Wuhan. (b) "Urban Catchments" Delegation in Nanjing. (c) M. Rybicki examines samples of biomonitoring (Source: UC Project)

toxicity and 2) the planning of decentralised wastewater management. The third stop was at Tongji University in Shanghai, where the goal was to exchange views on current decentralized wastewater management projects in the Chaohu region. As a result of this meeting, Shanghai Jiaotong University has become a new potential cooperation partner with ongoing projects in the Chaohu region. The subsequent visit allowed a detailed exchange of information about ongoing projects and the transfer of data from the project region. At the end of the trip, the Chenshan Botanical Garden (Shanghai) was visited, which is characterized by intensive research in the field of river water treatment (pollution by diffuse entries) and soil rehabilitation. The aim here was also the exploration of future cooperation. The next steps are the organization of workshops with the corresponding Chinese institutes in Germany as well as the initiation of a student exchange to consolidate the scientific relationship.

The BMBF-CLIENT project "Managing Water Resources for Urban Catchments" succeeded in building a first SME partnership: In the context of the 4th Governmental Consultations between Germany and PR China on 13.06.2016 in Beijing, AMC and WISUTEC as well as HC System and EWaters signed a Memorandum-of-Understanding for future cooperation between the German and Chinese SMEs (Fig. 2.13a–c). Their major interest is in building a joint venture for future cooperation in developing environmental information technology (including both soft- and hardware solutions) for sensor-based monitoring of environmental systems (e.g. for water supply and waste water). The company representatives as well as from Prof. Kolditz and Prof. Liao attended the 8th German Chinese Forum on Economic and Technology Cooperation hold at the same day and reported about ongoing projects in the Major-Water-Program as well as Sino-German cooperation concepts in future. The cooperation between research institution and companies – the so called "2+2 Concept" – shall foster research and development as well as implementation of novel environmental technologies.

Fig. 2.13 (**a**) Frank Neubert (CEO AMC), Cui Chen (UFZ), Olaf Kolditz (UFZ/TUD), Jonathan Fan (CEO HC System), Nicole Umlauf (BMBF Project Office Clean Water), Zhenliang Liao (Tongji University), Weijun Zhang (GM EWaters), Jan Richter (CEO GEOS/WISUTEC), and one colleague of (HC System) during the 8. German Chinese Forum on Economic and Technology Cooperation (from left to right). (**b**) J. Fan, F. Neubert, W. Zhang and J. Richter (from left to right) during the Signing Ceremony. (**c**) J. Fan, J. Richter, F. Neubert and W. Zhang (from left to right) after the Signing Ceremony (Source: UC Project)

The German companies AMC and WISUTEC met on 04.07.2016 in Shanghai with the Chinese companies HC System and Ewaters to coordinate cooperation in setting up a monitoring program for the city of Chaohu (Fig. 2.14). The talks were held in a very constructive atmosphere and were mainly used to inform each other about their own product lines and to identify interfaces for possible joint developments. The Sino-German consortium intends to apply for an Environmental Data Centre as part of a bid by the Asian Development Bank (ADB).

On 07–08.07.2016 the Digital Earth Summit took place in Beijing, which was hosted by the Chinese Academy of Sciences (Institute for Remote Sensing and Earth Observation). As part of the "Virtual Geographical Environments" (VGE) session, the "Virtual Environmental Information System - Chaohu" and the Helmholtz-CAS Network "Research Centre for Environmental Information Science" (RCEIS) were presented. The embedding of environmental information systems in virtual realities is of great interest both for scientific purposes (e.g. integration of large heterogeneous

Fig. 2.14 The presentations of the project partners served as information and to identify interfaces for possible joint developments (Source: UC Project)

Fig. 2.15 German delegation visited the Chao Lake Authority (Source: UC Project)

data sets) and decision makers for the visual support of complex planning processes. In the second part of the 3rd "Urban Catchments" trip in 2016, Prof. Kolditz, Prof. Liao, Prof. Kuang and Ms. Umlauf visited the Chao Lake Authority on July 12 and reported about the current progress of the BMBF-CLIENT project (Fig. 2.15). On

Fig. 2.16 (**a**) Visiting HC System in Shanghai. (**b**) O. Kolditz visited EWaters partner companies in Shanghai (Source: UC Project)

the way from Shanghai to Chaohu City, a stop was made in Wuxi. The Wuxi New District Construction Bureau has been briefed on the possibilities of environmental information systems. On the campus of Tongji University, the location for the urban measuring station was visited.

On 13.07.2016, Olaf Kolditz visited both HC System and EWaters partner companies in Shanghai (Fig. 2.16a, b). HC System, with more than 100 employees, is a growing expert for developing environmental system solutions in various areas. EWaters is a highly specialized consultant for water solutions planning based on their modelling expertise. Prof. Kolditz briefly reported about the progress of the "Managing Water Resources for Urban Catchments" project in Chaohu and particularly emphasized the related "Environmental Information System". However, comprehensive urban development includes much more than water resources and drinking water supply, i.e. domestic heating/cooling, energy supply ... (City4.0). Olaf Kolditz visited EWaters partner companies in Shanghai Virtual Information Systems (VIS) and admired the professional working level and the "young and dynamic" Chinese teams.

The first Chinese version of the WISUTEC software for the "Environmental Data Centre Chaohu" was developed in August 2016 and was set online (Fig. 2.17). It is planned to complete the final version by the beginning of 2017. The availability of environmental software in the national languages is an extremely important prerequisite for the acceptance of the research products of the BMBF project, especially among the authorities and stakeholders. With this software development, WISUTEC has not only made an essential contribution to the planned Environmental Information Center Chaohu, but also underlined the commitment of the German company partners to engage in the development of innovative environmental software (Water 4.0) in China in close cooperation with the Chinese company partners.

From 15 to 20.08.2016 Prof. Liao was a guest at CAWR at the TU Dresden (Fig. 2.18). The aim of the joint meeting was to discuss the status of both Chaohu's

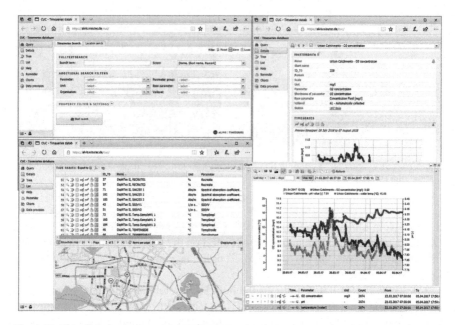

Fig. 2.17 First Chinese version of the WISUTEC software developed for the "Environmental Data Centre Chaohu" (Source: WISUTEC)

"Urban Catchments" and "Key Technologies and Management Modes for the Water Environmental Rehabilitation of a Lake City from the Catchment Viewpoint". Dr. Moldaenke from bbe has informed Prof. Liao about the new possibilities of biomonitoring.

During 17th and 19th of October 2016 the 2nd Sino-German Workshop of the Institutes of Hydrobiology (IHB) in Wuhan (China) and Dresden (Germany) took place in the Institute of Hydrobiology in Dresden (Fig. 2.19a, b). From IHB Wuhan Professor Dr. Hongzhu Wang, head of the Department of Freshwater Ecology, and Associate Professor Dr. Haijun Wang attended to the three-day workshop. After a short introduction of the IHB Dresden by Prof. Thomas U. Berendonk and three additional talks of institute staff, Prof. Hongzhu Wang talked about the current state of freshwater assessment in China using organisms (bioindicators) and current problems of aquatic ecosystems in China. Subsequently, Associate Prof. Haijun Wang explained the mesocosm facility of the Institute on the Lake Biandantong near Wuhan and the ongoing experiments, which focus on the effects of heavy nitrate pollution. The first day of the workshop was finalized by a visit of the biomonitoring field station of the IHB Dresden in the waste water treatment plant in Kreischa near Dresden. Dr. Marcus Rybicki explained the functioning of the station and the successful implementation and usage of the Daphnia-Toximeter from bbe Moldaenke. The second day of the workshop focused in two sessions on the enhancements and adjustment of the project ideas for the joint DFG-NSFC Call (Deutsche Forschungsgesellschaft

Fig. 2.18 Following the meeting with Z. Liao, M. Rybicki, D. Jungmann, C. Chen, O. Kolditz and C. Moldaenke met for a city walk around Dresden (Source: UC Project)

and National Natural Science Foundation of China) in 2017 and the second phase of the "Urban Catchments"-project. Prof. Berendonk and Dr.-Ing. Dirk Jungmann introduced the current project ideas and headed the following discussion. Dr. Bertram Boehrer from UFZ Magdeburg attended the session about the second "Urban Catchments" phase to complete the scientific partners of project part C. The workshop was finalized in the morning of the third day with the signature of a Memorandum of Understanding of both Institutes. A visit of our industry partners Dr. Frank Neubert (AMC) and Dr. Matthias Haase (WISUTEC) in Chemnitz accomplished this fruitful and discussion rich workshop.

On 18.10.2016, the Helmholtz Association - Chinese Academy of Sciences: "Strategic Partnership - Moving Ahead Together" Scientific Symposium took place in the institute of high energy physics (IHEP) of the Chinese Academy of Sciences (CAS) in Beijing (Fig. 2.20). After the opening greetings from Prof. Otmar D. Wiestler, the president of Helmholtz association and Prof. Tieniu Tan, the vice-president of Chinese Academy of Sciences, both sides have given presentations about the past, ongoing projects and future strategies. Prof. Kolditz presented the joint Sino-German Research Centre for Environmental Information Sciences (RCEIS) and its cooperation projects in China. Prof. Wiestler summarized the results of the bilateral Strategic Symposium. In addition, he proposed that China and Germany should open bilateral research programs in the following fields, such as geo- and environmental sciences, brain research, health and life sciences, energy materials and energy storage technology so that the outstanding scientists can exchange ideas in order to achieve

Fig. 2.19 (a) Discussion session in the seminar room of the Institute in Dresden. (b) M. Rybicki explains the functioning of the waste water treatment plant Kreischa and the position of the monitoring station (Source: M. Rybicki)

the win-win-based future-oriented strategic cooperation and the cooperation model should be guaranteed by joint research institutes.

On 21 October 2016, the delegation of the Helmholtz Association visited the Tongji University, in Shanghai. In the first session about "Information and data science", the president of Helmholtz Association, Prof. Wiestler introduced the Helmholtz strategy on information & data science. The vice-president of Tongji University, Prof. Zhiqiang Wu gave a talk about the innovation strategy of the university. Several talks in the area of smart manufacturing and industry 4.0 were given by the Chinese colleagues from Tongji University. Afterwards, the potential cooperation between Helmholtz association and Tongji University were discussed. Prof. Kolditz showed a video about the visualization of the environmental information

Fig. 2.20 Prof. Otmar D. Wiestler visits CAS-president Prof. Bai Chunli (Source: Helmholtz Association and Chinese Academy of Science)

system in Chao Lake, which is their current China-German cooperation project in China, funded by German ministry of Eduation and Research (BMBF). On the second session about "Clean water program", Prof. Xiaohu Dai, the Dean of College of Environmental Science and Engineering from Tongji University presented the Sino-German cooperation in Water Science, Technology and Education. Mrs. Nicole Umlauf, head of BMBF-Project "Clean Water" Office Shanghai, gave the talk about the BMBF water projects in China. Prof. Kolditz and Prof. Liao gave a joint presentation about the current state of the BMBF-CLIENT project "Managing Water Resources for Urban Catchments - Chaohu". At last, they emphasized the need to strengthen cooperation in the fields of water science, information technology and industry.

The Sino-German Major Water Programme Conference took place in Shanghai on 9th December, 2016 (Fig. 2.21). The Conference was to ensure the implementation of technical cooperation projects and promote the greater achievements of the Sino-German cooperation. The conference was organized by the Ministry of Science and Technology of PRC, Ministry of Environmental Protection, Ministry of Housing and Urban-Rural Development and the German Federal Ministry of Education and Research. As Chinese representatives, Dr. Chen Chuanghong, the Director General from Office of Major S&T Projects Most, and Mr. Liu Zhiquan, the General Inspector and Deputy Director General from the Department of S&T and Standards gave opening talks. As German representative, MinDirig Wilfried Kraus, the BMBF Deputy Director General gave an opening speech. During the conference, Prof. Wei Meng, who is an Academician and the technical team leader of Major Water Projects, talked about the progress of Major Water Program and Suggestions for Sino-German Cooperation. Dr. Christian Alecke from BMBF Division Resources and Sustainabil-

Fig. 2.21 The Sino-German Major Water Programme Conference in Shanghai (Source: UC Project)

ity introduced BMBF CLIENT Programme. In addition to that, the topics about Major Water Program in 13th Five-Year Plan such as the German prospects for future cooperation in Beijing-Tianjin-Hebei region and Tai lake Basin and system design for eutrophication control of Taihu lake, were also discussed. In the afternoon session, proposal for Major water projects of 13th Five-Year Plan based on existing projects and similar experiences were discussed. Prof. Xiaohu Dai from Tongji University talked about the progress of the Sino-German Major Water Programme. Prof. Kolditz, together with Prof. Liao, gave presentations about recommendations for monitoring, modelling, early warning programs and information systems. Prof. Andreas Tiehm from German Water Centre, together with Chinese partner, introduced the recommendations for drinking water safety, treatment and distribution.

2.4.3 Important Dates and Achievements in 2017

The German newspaper "Volksstimme" from Magdeburg reported on 14.02.17 on the transport of a research buoy from the Helmholtz Centre for Environmental Research UFZ in Magdeburg to Chao Lake, China (Fig. 2.22). Using the buoy, the researchers are able to investigate the algae infestation in the Chao Lake and thus contribute to the improvement of water quality. This was explained by the UFZ project manager

Fig. 2.22 German newspaper reported on the transport of a research buoy from Magdeburg to Chao Lake, China (Source: www.volksstimme.de)

for this area, Karsten Rinke from the Department of Lake Research of the "Urban Catchments"-project. With the transport of the research buoy to the Chao Lake, an important milestone for the project is within reach - the establishment of modern monitoring stations for monitoring water quality as an important element of the "Chaohu Early Warning System".

At the beginning of March 2017, the measuring buoy, which was jointly developed by engineering firm Plischke and the UFZ, arrived safely in Nanjing. Burkhard

Kuehn and Dr.-Ing. Marieke Frassl from the Department of Lake Research set up the measuring buoy on Chao Lake together with colleagues from the partner institute NIGLAS. After ensuring that no equipment had been damaged during delivery, Mr. Kuehn and Mrs. Frassl explained the procedure of the measuring buoy to her colleagues from NIGLAS: Jinge Zhu, Mr. Zeng Ye and Mr. Yun. In this practical experience, all language barriers were successfully overcome and the buoy was ready to go from the hardware side. Within a few days, the German partners Frank Neubert and Ziran Tao from AMC and their Chinese colleague Liu Yixiang from HC Systems jointly set up an FTP server, which is used for the online transmission of measurement data. On Sunday, the entire measuring system was successfully tested and put into operation. The buoy was then re-packed and delivered to Chao Lake at the beginning of the second week. Despite the bad weather the measuring buoy was successfully built within 1.5 days (Fig. 2.23a–c). The data set, measured by all sensors, is transmitted twice a day to the FTP server and can be checked online by NIGLAS colleagues. Of particular interest for the research are the data with the newly developed FluoroProbe from bbe Moldaenke. The data thus obtained provide a deeper insight into the development of blue-green carpets on Chao Lake. In the next few weeks, a database will be published by the German partners Matthias Haase and Markus Hillmann of WISUTEC to allow a simple analysis and quality control of the buoy data. All in all, the two weeks with testing and setting up the buoy was a great success, which strengthened the cooperation between the UFZ and NIGLAS.

Fig. 2.23 (a) Construction of the buoy. (b) Installation of all sensors of the buoy. (c) Successfully installed buoy on the Chao Lake (Source: UC Project)

Shortly thereafter, it was announced that HC Systems could provide a server with data ports for the transmission of buoy data. They have hosted the data of the buoy until the transmission was stopped by the Chinese side in July 2017.

On 27.03.2017 first results of the groundwater model of Chaohu were presented. Since the beginning of 2017, work has been ongoing on the Chaohu groundwater model, which is an integral part of the new "Work Package E" of the "Urban Catchments"-project. Surface waters are the major water resources in the North China Plain. However, subterranean water plays an important role in the water cycle and is a crucial transport path for pollutants entering surface waters - such as Chao Lake. The groundwater model is based on the available database in the region and the lessons learned from the ongoing project. Figure 2.24 shows an insight into the underground hydraulic system (groundwater levels) in the geographical context of the environmental information system - OpenGeoSys. The model will help to gain a better understanding of the interaction between the Chao Lake and the surrounding urban regions of Chaohu, as well as explain the occurrence of the algae toxin microcystin in groundwater.

On May 4, 2017, the installation of online data transmission from Chao Lake was successfully started. After the measuring buoy was anchored by the cooperation of the UFZ and the Chinese partner institute NIGLAS at Lake Chao, the buoy was able to start its measurements and the two German companies AMC and WISUTEC successfully put online data transmission into operation. Figure 2.25 shows the transmitted data during the test operation of the buoy ashore at the NIGLAS Institute. When setting up the software, it was possible to fall back on the technical support of the cooperation partner HC System, which provided the server technology in Shanghai. On the 21st of April the measured data could be researched and used by the Chinese operator of the buoy using the software AL.VIS/Timeseries. On the transmission path from the buoy to the server, the SensoMaster software is used, which transmits the measured data to the database server in a uniform protocol. Currently, almost 40 series of measurements are transmitted from the buoy. It is planned to connect further sensors for measurements in flowing waters to the information system.

Fig. 2.24 Visualization of the groundwater model of Chaohu (Source: UC Project)

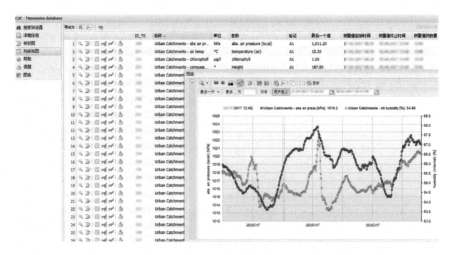

Fig. 2.25 Transmitted data during the test operation of the buoy (Source: WISUTEC)

Fig. 2.26 Participants of the 2nd "Urban Catchments" Status Conference in Leipzig (Source: UC Project)

On 13.06.2017 the 2nd "Urban Catchments" Status Conference took place in Leipzig. The UFZ project manager Prof. Kolditz organized this conference and the representatives of CAWR (UFZ, TUD) from the four SMEs (AMC, bbe, itwh, WISUTEC) and the OGS e.V. participated (Fig. 2.26). Each group presented their research results in the "Urban Catchments"-project, which was achieved in the last year. Mr. Helm and Mr. Wagner of WP-A (Urban Water Resources Management) gave presentations on "Urban Monitoring and Data Processing" and "Rainfall-Runoff Simulation Shuangqiao River". Afterwards, Mr. Li from the itwh presented the sewer network simulation of Chaohu City, focusing on data basis and usage, problems and proposed solutions. Mr. Aubron represented the WP-B (Decentralized Wastewater

Management) and reported on the latest results from the project entitled "Testing the Concept of Decentralized Wastewater Treatment in Chaohu Catchment". Miss Dr. Frassl and Mr. Rybicki (WP-C: Chao Lake) presented the results of hydrodynamic modelling at Chao Lake and informed about the planned biomonitoring system. In addition, Dr. Frassl installed in March with colleagues from the partner institute NIGLAS a measuring buoy on the Chao Lake. WP-D collated the information on the environmental information system. Mr. Pohl, the representative of the new WP-E (Groundwater System), explained the progress in groundwater modelling in the Chao Lake catchment. After the presentations the difficulties during the project and plans for the future were discussed.

From June 26 to July 8, 2017, Marcus Rybicki (WP-C) from the Institute of Hydrobiology of the TU Dresden (IHB-TUD) visited the Institute of Hydrobiology of the Chinese Academy of Sciences (CAS IHB) in Wuhan. After the successful transfer of the equipment for biomonitoring by the project partner bbe Moldaenke, the testing of the equipment and the installation of the biomonitoring at Bao'an Lake in Wuhan (field station) were on the program (Fig. 2.27a, b). During the first week of the two-week stay, bbe Moldaenke's daphnia toximeter and a selection of additional probes (IQSensorNet, WTW) for monitoring physico-chemical parameters were set up on a test bench in the laboratory. The CAS-IHB completed the selection of probes with a dedicated IQSensorNet probe for monitoring nitrate and ammonium since particularly high ammonium concentrations can lead to toxicity to aquatic invertebrates and fish. Furthermore, it was begun to optimize the cultures for daphnia and algae along with the staff and students of the institute in terms of quality control. Parallel to the laboratory work, a suitable location for the measuring station was established on the field station. In accordance with the requirements of the measuring location, the CAS-IHB started equipping the measuring station with a mains connection, an uninterruptible power supply and air conditioning. After completion of the laboratory test series, all equipment was transferred to the field station during the second week. Parallel to the ongoing optimization of the measurement site, the IQSensorNet was first put into operation and the probes were exposed in one of the field mesocosms of the field station. After installing the water pumps and laying the hoses, the daphnia toximeter was also put into operation. The first outdoor measurement series started on Wednesday, 05.07.2017. On July 6, the remote access to the devices via the mobile network was realized, so that the project partners of AMC could start with the configuration of the data transfer into the central project database. Based on the first measurement results of the daphnia toximeter, the adaptation of the configuration of the device to the measurement site has begun. On Friday, 07.07., a last joint maintenance of the daphnia toximeter and the transfer of the equipment to the Chinese colleagues took place. Finally, a list of tasks for the final optimization of the measuring station was prepared, which will be processed successively in the following days. Parallel to the practical work there was an intensive scientific exchange. So Mr. Rybicki gave a lecture on Friday, June 30, as part of the institute's "Freshwater Ecology Seminar", where he presented the CAWR and the Institute of Hydrobiology of the TU Dresden with its research focus. In the ensuing discussions, project ideas were concretized and opportunities for academic exchange, especially for students,

Fig. 2.27 (a) M. Rybicki with Master and Ph.D. student of Prof. Wang after the R workshop on 06.07.2017 in the premises of the "Daye Bao'an Lake Wetland Park". (b) H. Wang and M. Liu in front of the daphnia toximeter installed in the field station and Master Student C. Xu after successful installation of the IQSensorNet probes in a Pondmesocosmos (Source: UC Project)

were evaluated and planned. On Thursday, the 6th of July, due to the request of the Chinese colleagues, a workshop was organized and performed by Mr. Rybicki for the students of the working group headed by Prof. Hongzhu Wang for the introduction into the statistical and development environment R.

In July 2017, the delegation led by Saxon State Ministry of Environment and Agriculture (SMUL) Thomas Schmidt visited Tongji University (Fig. 2.28). Tongji University cooperates intensively with the Leipzig-based Helmholtz Centre for

Fig. 2.28 Minister of State Thomas Schmidt and his delegation visited Tongji University in May 2017 (Source: UC Project)

Fig. 2.29 In an intensive course, fundamentals of environmental fluid mechanics were imparted by M. Walther and O. Kolditz (Source: UC Project)

Environmental Research in the "Urban Catchments"-project and is working on the development of an environmental information system as the basis for restructuring strategies for the Chao Lake. The project, which is jointly managed by the UFZ and the TU Dresden under the consolidation of the CAWR (Centre for Advanced Water Research), is the largest German-Chinese environmental research project in Anhui Province.

After the delegation visit of the Saxon Minister of State Thomas Schmidt, JProf. Walther (Pollutant Hydrology at TU Dresden) and Prof. Kolditz accepted the invitation by Prof. Liao at the Tongji University for giving lectures (Fig. 2.29). In an intensive course, fundamentals of environmental fluid mechanics were imparted. The Tongji University and the TU Dresden want to deepen their joint activities not only in research but also in teaching. This is more than obvious - citing the unity of research and teaching - because both universities are involved in UNEP programs. The UN Environment-Tongji Institute for Sustainable Development (ISED) was established in 2002.[4] The post-graduate training program CIPSEM initiated by UNEP, UNESCO and BMUB has been implemented at the TU Dresden since 1977.[5] The target groups for both programs are specialists and executives from developing and emerging countries working in the environmental field. TU Dresden and the Helmholtz Centre for Environmental Research–UFZ, which founded the joint CAWR water centre,[6] together with Tongji University and other Chinese partners, want to teach along the Yangtze River to deepen their understanding of the complex and diverse environmental processes in river basins to students and professionals (RIVERCHALLENGE - "From Source to the Sea"). On the sidelines of the courses there was also opportunity with Profs. Dai and Dohmann to talk about current developments in the "Major Water Program".

From 21 to 25.07.2017 Prof. Liao and Mr. Fan (CEO) and Li from HC System visited the project partners of "Urban Catchments" (Figs. 2.30 and 2.31a), Chemnitz (Fig. 2.31b) and Dresden (Figs. 2.32, 2.33 and 2.34). The aim of the trip was above all the visit of the German company partners AMC, WISUTEC and itwh with the demonstration of UC products. HC Systems was able to convince itself of the performance of the system solutions on site and thus received further important information for the application for the ADB project Chaohu for the implementation of an environmental information system. At the visualization centre of the UFZ in Leipzig, the virtual environmental information system Chaohu was presented by Dr. Karsten Rink and Lars Bilke. In particular, concepts and possibilities of integrating complex, heterogeneous, large data sets using the OGS DataExplorer in a uniform geographical context were demonstrated. In addition, process models for different aquatic compartments can be integrated. Other possibilities for using VR systems, e.g. for geothermal deposits, were demonstrated by JProf. Dr. Haibing Shao. Another point was the presentation of the book project "Urban Catchments - Chaohu" by Dr. Agnes Sachse and the concrete involvement of the Chinese side in the publication project. In Chemnitz, the guests discussed with the company representatives of AMC and WISUTEC the status of the work for the connection of measuring systems to the server at HC System and for the research by means of the implemented software for the data center Chaohu. In addition, an even stronger cooperation between IT specialists from the three companies HC System, AMC and WISUTEC was agreed. The aim is to be able to develop a competitive offer for the ADB project "Data Center Chaohu". In addition to the visit to Chemnitz, the visit of the test stations (TUD) on the sewage treatment plant in Kreischa followed. Until the organizational handling of the transport of the measuring stations, the technical modalities for the integration of

---

[4] http://unep-iesd.tongji.edu.cn.
[5] https://tu-dresden.de/bu/umwelt/cipsem.
[6] https://www.ufz.de/cawr.

Fig. 2.30 Demonstration of the virtual environmental information system Chaohu in the Vislab of the UFZ in Leipzig (Source: UC Project)

Fig. 2.31 (**a**) The delegation around Prof. Liao visited the UFZ in Leipzig and WISUTEC in Chemnitz (**b**). (Source: UC Project)

further third-party measuring systems into the current server system of HC System, AMC and WISUTEC could be clarified on site. The future possible data acquisition for the provision of water quality parameters and the required hydrographs for model calibration were demonstrated. Based on the objective of the project "Urban Catchments" for the development of a comprehensive urban water management and environmental information system to increase the sustainable, rising water quality of the Chaohu region, further possibilities for existing system solutions in the Dresden area were demonstrated. Realistic solutions for the strategy-controlled real-time storage management in existing channel systems were presented. The central control system over the entire drainage system of the city of Dresden was set up in the sewage treatment plant "Kaditz" and was presented to the delegation.

From 18.08. until 02.09.2017 Benjamin Wagner (ISI-TUD, member of the WP-A) visited the project partners of "Urban Catchments" in Shanghai. The aim

Fig. 2.32 Visit of the sewage treatment plant Kreischa near Dresden (Source: UC Project)

Fig. 2.33 Visit to the sewage treatment plant in Kaditz and the control system implemented by itwh (Source: UC Project)

of the trip was above all the scientific exchange of the project partners for WP-A. Furthermore, the journey served as preparation for the dispatch and integration of the monitoring stations into the environmental information system at HC Systems. At Tongji University's College of Environmental Science and Engineering, the model of the Shuangqiao River Basin was discussed with Yufeng Lou (Fig. 2.35). Afterwards it was possible to coordinate with Chongjia Luo the important points for the setup, the test operation and the preparation for the integration of the monitoring stations. In a first test operation measurement data from Germany were sent to the HC Systems server. With the help of these tests, any errors in the transmission system are detected and quickly resolved. Through energetic support from Prof. Liao, important boundary conditions for transport and customs modalities could be clarified. A site survey was

Fig. 2.34 View into the machine hall of the sewage treatment plant Kaditz (Source: UC Project)

Fig. 2.35 B. Wagner (ISI-TUD) visited the project partners of "Urban Catchments" in Shanghai, Prof. Liao (Tongji University) (Source: B. Wagner)

conducted to check the information obtained by satellite imagery about important structural structures in the flow cross section of the Shuangqiao River, which have an influence on the discharge behaviour in heavy rainfall events (Fig. 2.36). Any special features (e.g. bridge piers) can thus be excluded or confirmed. Other important input parameters (e.g. soil infiltration capability, special climatic boundary conditions) were discussed and evaluated with the help of the Chinese partners.

On November 17, 2017, a signing ceremony took place in Wuhan on the occasion of the 10th anniversary of the regional partnership between the Free State of Saxony and the province of Hubei (Fig. 2.37a). "Urban Catchments" has actively participated in the delegation trip of the Free State of Saxony. During the ceremony, the directors of the Institute of Hydrobiology of the Chinese Academy of Sciences CAS-IHB (Prof. Hongzhu Wang) and the Technical University of Dresden (Prof. Thomas Berendonk) completed a cooperation agreement (Fig. 2.37b). The latter agrees to further intensify the scientific cooperation of the hydrobiologists of both countries, and the MoU also supports the submission of further joint research projects, such as "Urban Catchments - Wuhan". In addition, cooperation in education is also sought. At the technology forum in the context of the delegation trip on November 16, 2017 Prof. Kolditz presented the ongoing "Urban Catchments"-project Chaohu. In addition, Prof. Berendonk presented the Institute of Hydrobiology and the "Center for Advanced Water Research - CAWR" in the seminar of the CAS Institute of Hydrobiology. During the delegation trip there was also a conversation with Minister Li (MEP Hubei). His return visit to Dresden took place in December.

Organized by WP-A project, four online water monitoring stations arrived at the campus of Tongji University in Shanghai at the beginning of December 2017. Christian Koch from the Urban Water Management of the TU Dresden was on site to coordinate unloading, set-up and commissioning. Together with Professor Liao and especially his students, all measuring stations were unpacked and all the respective sensors connected. The system offers a wide range of parameters, in addition to a number of physico-chemical parameters, the water level and flow of the channel or flowing water can be detected. After calibrating most of the sensors, a measuring station was installed at a wastewater pumping station on the campus (Fig. 2.38). Above all, this procedure serves the experience gained by the operators on site. Thus, data transmission, maintenance and calibration routines can be trained for later use of the measuring stations in the test area. In spring 2018, the measuring systems will be relocated to the Chao Lake, allowing measurement data to be collected, in particular, for the calibration of the already created numerical models.

Fig. 2.36 Site visit in Chaohu (Source: B. Wagner)

Fig. 2.37 (a) Signing of the Cooperation Agreement between the Directors of the Institutes of Hydrobiology of the Chinese Academy of Sciences CAS-IHB (Prof. Hongzhu Wang) and the Technical University of Dresden. (b) Discussions on the joint contract (Source: UC Project)

Fig. 2.38 The online monitoring stations were successfully installed in Shanghai (Source: Tongji University)

2.5 Project Partners

Due to its complexity, this research project is interdisciplinary and therefore relies on the support of numerous partners. This section gives an overview of all involved Chinese partners from science, administration, and government agencies as well as the German project partners from science and industry.

2.5.1 Chinese Partners

2.5.1.1 Tongji University

Tongji University, formerly Tongji German Medical School, was founded by Erich Paulun, a German doctor in 1907. The name Tongji suggests cooperating by riding the same boat. It was one of the oldest and most prestigious institutions of higher education in China. The university has developed rapidly in all respects since the country's opening-up policy in 1978. As one of the leading universities in China, it is now a comprehensive university with ten major disciplines in sciences, engineering, medicine, humanities, law, economics, management, philosophy, arts and pedagogy with strength in architecture, civil engineering and oceanography. The UNEP-Tongji Institute of Environment for Sustainable Development (IESD) was jointly established by United Nations Environment Programme (UNEP) and Tongji University (Tongji).

The College of Environmental Science and Engineering (CESE) at Tongji University was one of the earliest established colleges dedicated to environmental education and research in China. Environmental Engineering is one of the national key disciplines. During the past decades, Tongji University makes great efforts to research and practice for pollution control and improvement of water quality in the Chaohu area, and gave its special contributions to the water environmental rehabilitation of Chaohu.

2.5.1.2 Chinese Academy of Science

The Chinese academy of Sciences (CAS) was established on November 1, 1949, in Beijing, where it is headquartered. It was formed from several existing scientific institutes and soon welcomed over 200 returning scientists who contributed to CAS the high-level expertise they had acquired abroad. It is the world's largest research organisation, comprising around 60,000 researchers working in 114 institutes and has been consistently ranked among the top research organisations around the world.

2.5.1.3 Chinese Academy of Science: Institute of Hydrobiology

With a history of 82 years, Institute of Hydrobiology (hereinafter abbreviated as IHB), Chinese Academy of Sciences (hereinafter abbreviated as CAS), is a comprehensive academic research institution which devotes to the studies of life processes of inland aquatic organisms, ecological environment protection and utilization of biological resources. It was evolved from Natural History Museum of Academia Sinica founded in Nanjing, January 1930, and was renamed in July 1934 as Institute of Zoology and Botany of Academia Sinica. The Institute was divided into two in May 1944, i.e. Institute of Zoology and Institute of Botany under Academia Sinica. In February 1950, after the establishment of CAS, the main division of Institute of Zoology of Academia Sinica, Institute of Botany of Academia Sinica, the division of phycology of Shandong University and some researchers from the National Academy of Peking were merged into Institute of Hydrobiology, Chinese Academy of Sciences. It was then located in Shanghai, but moved to Wuhan in September 1954. In 2011, IHB entered the pilot project of the CAS Innovation 2020 programme. In 2015, it became one of the feature institutes in the CAS pioneer initiative.

2.5.1.4 Chinese Academy of Sciences: Nanjing Institute of Geography and Limnology

Nanjing Institute of Geography and Limnology (NIGLAS), Chinese Academy of Sciences, formerly Geography Institute of China, was founded in Beibei, Chongqing in August 1940. It is the only institute specializing in the research of lake-basin system in China and was ever directed by Huang Bingwei, Ren Meie, and Zhou Lisan who are Chinese Academy of Sciences (CAS) Academician at different time in its history. Research fields in NIGLAS include:

- lake environment protection and resources utilization
- lake-basin system evolution and manipulation
- regional sustainable development with focus on lake sediment and environment evolution
- lake hydrology and water resources
- lake biology and ecology
- lake environment and engineering
- lake-basin process and manipulation
- resources and environment of basin and regional development
- lake-basin monitoring and digital basin

2.5.1.5 HC Systems

HC Systems,[7] with more than 100 employees, is a well-known digital technology service company and senior intelligence expert in the environmental field. Founded in 2005, the company specializes in the automation, information, and intelligent high-tech services for industrial control processes in the environmental field. The business scope covers various water environmental industry chains such as municipal water treatment (water supply and drainage), pipe network pumping stations, urban flood control, solid waste treatment, etc.; integrated system research and development, smart water supply, water environment treatment solutions, engineering construction.

2.5.1.6 EWaters

EWaters Corporation[8] specializes in consulting and environmental technology transfer into the Chinese market, with a main business focus on advanced technological applications in all water systems. Staying at the leading edge of modern modelling technology, risk management, water quality monitoring technology and asset management, the accumulated wealth of local and international expertise enables to provide innovative, intelligent and integrated solutions and to help the clients to achieve the best management practice.

2.5.1.7 Chao Lake Management Authority

The Chao Lake Management Authority (CLMA) of Anhui province was established in 2012. CLMA focuses on a number of tasks about Chao Lake including planning, environmental protection and tourism for the comprehensive planning, governance, development and application of Chao Lake.

---

[7]http://www.haocang.com/.
[8]http://www.ewaters.biz/.

2.5.2 German Partners

2.5.2.1 Helmholtz Centre for Environmental Research

The Helmholtz Center for Environmental Research–UFZ was founded in 1991 under the name UFZ-Umweltforschungszentrum Leipzig-Halle GmbH and has more than 1100 employees at its locations in Leipzig, Halle and Magdeburg. It investigates the complex interactions between man and the environment in used and disturbed landscapes, especially densely populated urban and industrial agglomerations and natural landscapes. The scientists of the UFZ develop concepts and procedures that should help to ensure the natural foundations of life for future generations.

Department of Environmental Informatics

The Department of Environmental Informatics at the UFZ in Leipzig in cooperation with the Chair of Applied Environmental Systems Analysis at the TU Dresden is developing methods and software for the simulation of terrestrial environmental processes. As a result, the OpenGeoSys (OGS[9]) software is being developed for over 25 years. It offers a multitude of algorithms for the numerical simulation of coupled thermo-hydro-mechanical-chemical (THMC) processes in porous media and can be applied for numerous hydrological and geotechnical applications (Kolditz 2002; Kolditz et al. 2012). In addition to the numerical process simulation with OGS THMC, data processing workflows are developed for environmental system analysis. This includes data integration from various sources (e.g. remote sensing, surface monitoring, geological data) or the visualization of complex data collections for decision making processes based on simulation results from different scenarios. An important instrument for this is the OGS Data Explorer (Rink et al. 2013, 2014). The Department has numerous collaborations with the People's Republic of China, with projects focusing on environmental challenges and energy storage applications. Within the "Urban Catchments"-project, members of the department developed a Virtual Geographic Environment (VGE) for the exploration and understanding of complex collections of geoscientific and numerical data sets.

Environmental and Biotechnology Centre

The UFZ department "Environmental and Biotechnology Centre" (UBZ) has been developing, testing and implementing decentralised, nature-based (green) technologies and management concepts for (waste)water management since 2002. In addition to new and improved green (waste)water technologies, innovative operational and organisational models have been developed to ensure sustainable and efficient resources management. In addition, the research carried out by the UBZ aims at creating or adapting legal frameworks for the successful implementation of technical and managerial solutions in close cooperation with decision-makers. Relevant project examples that were managed by UBZ or where the UBZ took a leading role

---

[9]www.opengeosys.org.

include: (1) Urban Transformation: Feasibility Study to Model Integrative Urban Redevelopment, Including Wastewater, Energy and Waste (UBA); (2) SWINGS: Safeguarding Water Resources in India with Green and Sustainable Technologies (EU); (3) NICE I and II: Formation, Conception and Moderation of a National Committee for the Implementation of Decentralized Wastewater Management Scenarios in Jordan (BMBF); (4) SMART I and II: Integrated Water Resources Management in the Lower Jordan Rift Valley (BMBF); and (5) SMART-Move: Management of Highly Variable Water Resources in Semi-Arid Regions (BMBF).

Department of Lake Research

The Department of Lake Research (SEEFO) at the UFZ in Magdeburg consists of an interdisciplinary team of 9 scientists, 7 technicians and a varying number of third-party funded scientists. The strength of the scientific team is the process-oriented handling of complex questions in the aquatic area. Extensive experience from program-oriented research (POF) as well as from numerous third-party funded projects exists in the areas of lake and reservoir management, lake restoration, water monitoring, aquatic biogeochemistry, and lake modeling. Special skills are available in the fields of limnophysics, biogeochemistry, microbiology, plankton ecology and modeling. Current priorities in ongoing research in the Department SEEFO are on the online water quality monitoring, coupled physical-ecological modeling, and the carbon cycle in lakes and reservoirs. International research activities of the department are focused on South America, China, the Caucasian area as well as Europe. As part of the "Urban Catchments"-project, the Department of Lake Research is involved in the development, construction and evaluation of the online monitoring system of the Chao Lake.

2.5.2.2 Technische Universität Dresden

The Technische Universität Dresden (TUD) is one of eleven German universities that currently have the status of a university of excellence. It is the largest university in the federal state of Saxony.

Institute of Urban and Industrial Water Management

The Institute of Urban and Industrial Water Management (TUD-SWW) has 39 employees in the chairs of Urban Water Management, Industrial Water Management and Drinking Water Supply (03/2018). In the field of wastewater treatment, the employees of the Chair of Urban Water Management deal with processes and systems of municipal and industrial wastewater and sludge treatment as well as their optimization. In the field of urban drainage, the main topics are the integrated consideration of urban drainage systems for holistic optimization and immission-oriented requirements, the cost-efficient development of urban drainage systems, the investigation of transport and sales processes in sewers and urban waters as well as the adaptation of urban water management to changing framework conditions. The professorship has extensive experience in coordinating and carrying out research activities of projects funded by public funds (e.g. EU, BMBF, BMWi, German Federal Environmental Foundation, German Research Foundation) but also in the field of contract research the employees have many years of experience. The operation and the evaluation of online monitoring systems of the water quantity and quality in the sewage system and in urban waters have been tested in numerous national and international campaigns. Since 2012, an observatory for the long-term observation and analysis of drainage processes is under construction in Dresden. The findings from this preliminary work flow directly into the capacity development of the Chinese cooperation partners as well as the simplification of the measuring stations via proxy measurements. In cooperation with AMC and WISUTEC, the integration of data into a server-based management system was tested. The data provide an indispensable basis for understanding the pollution processes, optimizing the effluent system efficiently, and minimizing urban flooding. The IWAS (International Water Research Alliance Saxony) project has provided experience in working in fast-growing urban systems in Brasilia. The aspect of changing framework conditions for urban water management planning could also be methodologically and substantively examined in other IWAS regions as well as in several national research projects. The proposed two-step procedure for the identification and minimization of the water pollution could already be applied successfully in several times. The resulting combinations of activities form the basis for integrated development and management concepts of urban water management. As part of the "Urban Catchments"-project, the Chair of Urban Water Management has developed a system for the development of an online monitoring system for water quality in Chaohu and in the Chao Lake. This includes the monitoring of urban wastewater and water networks.

Institute of Hydrobiology

The Institute of Hydrobiology at TU Dresden (TUD-HYB) has demonstrated its expertise in biological effect monitoring in numerous publications and projects in recent years (Mazurova et al. 2010; Jungmann et al. 2001; Jungmann et al. 2004, 2009; Ladewig et al. 2006; BMBF: IWRM Mozambique and KoPiGe, IWAS Scheifhacken et al. 2011; Ertel et al. 2012; Seiler and Berendonk 2012). In addition to monitoring, the focus was also on risk assessment. The development and evaluation of test strategies for the determination of an environmental risk and the establishment of standard test procedures for the OECD have been part of applied research at the institute for many years (Federal Environment Agency FKZ: 295 63 075; FKZ: 299 65 221/05FKZ: 360 12; FKZ: 202 67 437; FKZ: 299 67 44). The personnel and technical capacities at the Institute of Hydrobiology include 3 scientific and 4 technical staff, 1 administrative employee, 2 post-doctoral student, 6 doctoral students and 4 Bachelor and 6 Master students. In the evaluation of long-term monitoring data, the institute has extensive experience from projects with dams and streams (BMBF KoPiGe). As part of an international project in Mozambique and Namibia, a biomonitoring concept has been developed for the city of Tete, which lies within the coal mining industry at the mining area on the Zambezi, and scientists from Mozambique qualified. Together with the water authority ARAZambezi, a report was prepared on the effects of the river depression on the Sambesi fauna. The Institute was responsible for the coordination of research in Ukraine within the IWAS (International Water Science Alliance Saxony) Initiative (BMBF) for concepts for the efficient improvement of surface water quality. Within the "Urban Catchments"-project, a field study at the waste water treatment plant Kreischa (near Dresden, Germany) was performed to gain experience with the Daphnia-Toximeter and the FluoSens under field conditions and to apply this technology to the Chao Lake, China.

Institute for Groundwater Management

The Institute for Groundwater Management (TUD-IGW) consists of the Professorship of Groundwater Management (Prof. Dr. R. Liedl) and the Junior-Professorship of Contaminant Hydrology (JProf. Dr. M. Walther). The institute was founded in the year 1946 and emerged from the former Institute of Soil and Water Management and is since than an essential part of the Department Hydrosciences at the Technische Universität Dresden. The IGW provides a broad educational background to Bachelor and Master studies focussing on various methodological and applied topics. Scientifically, the institute aims to address both, fundamental and applied research questions. Examplary current working topics enclose stable isotope chemistry, tracer technologies, karst hydraulics, coastal groundwater systems or aquifer storage and recovery systems. Thereby, the institute provides expertise in the application and development of numerical models together with respective field monitoring and laboratory analysis.

2.5.2.3 AMC–Analytik & Messtechnik GmbH Chemnitz

AMC–Analytik & Messtechnik GmbH Chemnitz[10] is an SME with more than 20 years of experience in the fields of monitoring and process control systems, measuring, testing and information systems. More than 20 engineering graduates, technicians and skilled workers guarantee the development of complete project solutions through the phases of planning and configuration, programming, installation and commissioning, documentation and training as well as maintenance and servicing. Qualifications in the fields of electrical engineering, computer science, automation technology and process engineering form the basis for a detailed analysis of the technical, technological and physical processes as a basis for the development of complete system solutions. Customers include German and foreign companies and institutions from a wide range of industries, such as Public sector clients, power engineering, automotive industry, mechanical and plant engineering, primary industry as well as research/development and training. Based on market-proven industrial components, AMC develops tailor-made solutions ranging from signal acquisition in hydrological monitoring and early warning stations to centralized and decentralized processing to distributed output and evaluation of data in local networks and on

---

[10] http://www.amc-systeme.de.

the Internet. For the "Urban Catchments"-project, AMC implemented the complete chain of an environmental information system.

2.5.2.4 WISUTEC Umwelttechnik GmbH

The administration of large, inhomogeneous and spatially distributed data covering a variety of different topics is one of the challenges of working on major environmental or cleanup projects today WISUTEC[11] experience in this area comes from working on the large Wismut GmbH project. During years of working on one the most important environmental projects in Germany and Europe - the remediation of the residues of uranium ore mining in Saxony and Thuringia - WISUTEC was able to develop the web-based AL.VIS information system with its research, analysis and GIS/map features. WISUTEC Umwelttechnik GmbH has extensive experience in the planning and implementation of environmental monitoring systems in Germany, Eastern Europe, Russia and Central Asia. The company develops web-based database solutions with GIS functionalities for national and international projects. This includes information systems for the management of:

- environmental data (surface water, groundwater, climate values, substance samples),
- GIS information,
- and data and information about the observed objects, e.g. documents, photos, maps.

The range of services also includes:

- Application programming (.NET, C#), especially in the area of environmental data management
- Database development (ORACLE, MS SQL Server, PostgreSQL/PostGIS)
- Implementation and hosting of (web) portals
- Development of IT Concepts
- IT consulting services
- Data migration ("legacy data transfers")

In recent years, WISUTEC has participated in several projects for the development of early warning systems in the field of environmental monitoring of waters. With the practical experience gained, important work was carried out on the development of software modules for an early-warning system in the R&D project "Conception and exemplary implementation of a pilot station for water body monitoring with regard to radiological and chemical-toxic ingredients". In the "Urban Catchments"-project, a comprehensive environmental information system was set up.

---

[11] http://www.wisutec.de/en-gb/Company.

2.5.2.5 bbe Moldaenke GmbH

For more than 20 years bbe Moldaenke GmbH[12] has been one of the leading manufacturers of environmental technology products. bbe develops and produces measuring instruments and software for water quality control. bbe devices are used in oceanography and limnology, drinking water quality control, raw water control, bath water quality assessment, aquaculture system monitoring and environmental control. bbe Moldaenke GmbH specializes in the construction and development of spectrofluorometers for chlorophyll analysis and for the analysis of naturally occurring pigments (algae pigments, etc). In addition, bbe Moldaenke GmbH is the market leader in the field of biological early warning systems, the toximeters/biomonitors for the detection of environmentally harmful substances and mixtures in drinking water intake and in environmental monitoring. bbe has been building and selling Daphnia algae and fish biomonitors for about 20 years. However, sensitive monitors are not straightforward, which means that adjustments have to be made over and over again. For this purpose, proprietary systems have been developed which are used worldwide (eg China). The bbe team focuses on important tasks such as development, quality control and customer care. Many simple works of the production process have therefore been outsourced. International cooperation requires presence in many places simultaneously outside of Germany (more than 35 countries). bbe has been building spectrofluorometers for 25 years. The aim of this device is the humic substance analysis and the evaluation of the elimination of high molecular weight fractions of humic substances. It is believed that high molecular weight fractions contribute to the growth of bacteria in drinking water, so the flocculation in a waterworks can be controlled according to this parameter. This aspect was very interesting within the "Urban Catchments"-project regarding waters with high sewage and agricultural runoff content, such as Chao Lake.

2.5.2.6 Institute for Technical-Scientific Hydrology GmbH

The institute for technical-scientific hydrology GmbH (itwh)[13] with headquarters in Hanover and branches in Dresden, Flensburg and Nuremberg was founded in 1987 and has more than 70 employees, mostly engineers, hydrologists and computer scientists. The main activities of itwh lies in the creation of studies, concepts, R&D projects, etc. in urban areas such as drainage concepts, general drainage plans, urban flooding studies, concepts and measures of rainwater management, realtime control of sewer systems, etc. As great successful projects can be mentioned: Changde, Chaohu (PRC); Dresden (Fuchs et al. 2015), Düsseldorf, Frankfurt, Hamburg, Hannover, Karlsruhe (Germany); Hanoi (Vietnam); Milwaukee (USA)[14]; Warsaw (Poland); Vienna (Austria) (Fuchs and Beeneken 2005). In the software area,

---

[12] https://www.bbe-moldaenke.de/en/.
[13] https://www.itwh.de/en/.
[14] https://www.itwh.de/files/dokumente/referenzen/Projekt_RTC-Milwaukee.pdf.

simulation models and GIS applications for the rainfall-runoff transport process in the urban area are being developed. The itwh operates predominantly in Germany and the neighboring states, in cooperation with other engineering offices/universities but also in China and Vietnam. In the Chinese city of Changde/Hunan Province, an ecologically oriented master plan for the urban areas as well as the adjoining surrounding area was set up, in which numerous measures of rainwater management and water remediation are conceived and partly developed. In Chaohu, a master plan was drawn up with the aim of developing the city into an ecological seaside city.[15] For the urban catchment project itwh was responsible for the modelling and planning of the rainwater management and sewer systems of Chaohu.

2.5.2.7 OpenGeoSys e.V.

Modelling and computer-aided simulation in the field of environment and geotechnical engineering is becoming increasingly important in decision-making at the municipal, national and international level. The results of such computer models often act as a bridge to the knowledge transfer between science, business, politics and population. This is especially the case if the processes take place underground and are thus usually hidden from the eye of the observer, such as e.g. in case of soil contamination or geothermal energy production. The aim of the German non-profit association OpenGeoSys e.V. is to create better transparency and in particular to inform about how such models and concepts are responsibly developed, created and subsequently communicated, so that decision-makers can evaluate them and the media and the population can understand them. The OGS e.V. helps to shape the communication between science, research, politics, administration, companies, as well as media and public in the environmental and geotechnical research field. A main focus is on the coordination of knowledge and technology transfer and cooperation between university and non-university institutions. This also includes the promotion of specialist qualification in the field of environmental technology in the context of training and further education measures.

References

Ertel A, Lupo A, Scheifhacken N, Bodnarchuk T, Manturova O, Berendonk T, and Petzoldt T. Heavy load and high potential. Anthropogenic pressures and their impacts on the water quality along a lowland river (Western Bug, Ukraine). Environ. Earth Sci. 65(5), 1459–1473 (2012)

Fuchs L, and Beeneken T. Development and implementation of a real-time control strategy for the sewer system of the city of Vienna. Water Sci. Technol. 52(5), 187–194 (2005)

Fuchs L, Krebs P, Lindenberg M, Männig F, and Seggelke K. Weiterentwicklung des Entwässerungskonzepts der Stadt Dresden vor dem Hintergrund klimatischer Veränderungen. Korrespondenz Wasserwirtschaft, 8 (2015)

---

[15] https://www.wasser-hannover.de/de/projekte/oekologische-seestadt-chaohu.

Jungmann D, Brust K, Hultsch V, Licht O, Mahlmann J, Schmidt J, and Nagel R. Significance of test on the ecosystem level for the risk assessment of Hazard compounds in surface water, Part II: impact of the herbicide terbutryn (Umweltbundesamt, 2001), p. 123

Jungmann D, Ladewig V, Ludwichowski KU, Petzsch P, and Nagel R. Intersexuality in Gammarus Fossarum Koch - a Common Inducible Phenomenon? Archiv Fur Hydrobiologie **159**, 511–529 (2004)

Jungmann D, Bandow C, Gildemeister T, Nagel R, Preuss TG, Ratte HT, Shinn C, Weltje L, and Maes HM. Chronic Toxicity of Fenoxycarb to the Midge Chironomus Riparius after Exposure in Sediments of Different Composition. J. Soils Sediments **9**, 94–102 (2009)

Kolditz O. *Computational Methods in Environmental Fluid Mechanics*. Springer Science & Business Media (2002)

Kolditz O et al. OpenGeoSys: an open-source initiative for numerical simulation of thermo-hydro-mechanical/chemical (THM/C) processes in porous media. Environ. Earth Sci. **67**(2), 589–599 (2012)

Krüger T, Wiegand C, Kun L, Luckas B, and Pflugmacher S. More and more toxins around-analysis of cyanobacterial strains isolated from Lake Chao (Anhui Province, China). Toxicon **56**(8), 1520–1524 (2010)

Krüger T, Hölzel N, and Luckas B. Influence of cultivation parameters on growth and microcystin production of Microcystis aeruginosa (Cyanophyceae) isolated from Lake Chao (China). Microb. Ecol. **63**(1), 199–209 (2012)

Ladewig V, Jungmann D, Kohler HR, Schirling M, Triebskorn R, and Nagel R. Population structure and dynamics of Gam-marus fossarum (Amphipoda) upstream and downstream from effluents of sewage treatment plants. Arch. Environ. Contam. Toxicol. **50**(3), 370–383 (2006)

Mazurova E, Hilscherova K, Sidlova-Stepankova T, Kohler HR, Triebskorn R, Jungmann D, Giesy JP, and Blaha L. Chronic toxicity of contaminated sediments on reproduction and histopathology of the crustacean Gammarus fossarum and relation-ship with the chemical contamination and in vitro effects. J. Soils Sediments **10**(3), 423–433 (2010)

Rink K, Fischer T, Selle B, and Kolditz O. A data exploration framework for validation and setup of hydrological models. Environ. Earth Sci. **69**(2), 469–477 (2013). https://doi.org/10.1007/s12665-012-2030-3

Rink K, Bilke L, and Kolditz O. Visualisation strategies for environmental modelling data. Environ. Earth Sci. **72**(10), 3857–3868 (2014). ISSN 1866-6299. https://doi.org/10.1007/s12665-013-2970-2.

Scheifhacken N, Haase U, Gram-Radu L, Kozovyi R, and Berendonk TU. How to assess hydromorphology? A com-parison of Ukrainian and German approaches. Environ. Earth Sci. **65**(5), 1483–1499 (2011)

Seiler C, and Berendonk TU. Heavy metal driven co-selection of antibiotic resistance in soil and water bodies impacted by agriculture and aquaculture. Frontiers in Microbiology **3**, 399–399 (2012)

Chapter 3
WP-A: Urban Water Resources Management

Peter Krebs, Firas Al Janabi, Björn Helm, Honghao Li, Benjamin Wagner, Christian Koch, Renyuan Wang and Lothar Fuchs

3.1 Introduction

The Anhui Chao Lake Environmental Rehabilitation Project (ACLERP) calls for the laying of 100 km of sewer and reaching 30,000 m^3/d in capacity at the wastewater treatment plant. The "Masterplan Ökologische Seestadt Chaohu"[1] names the remaining deficits of ACLERP and gives concrete recommendations for further plans.

---

[1] https://www.wasser-hannover.de/de/projekte/oekologische-seestadt-chaohu.

P. Krebs
Department of Hydrosciences, Institute of Urban and Industrial Water Management,
Technische Universität Dresden, Bergstraße 66, 01069 Dresden, Germany
e-mail: peter.krebs@tu-dresden.de

F. Al Janabi · B. Helm · B. Wagner (✉) · C. Koch
Department of Hydrosciences, Institute for Urban and Industrial Water Management,
Chair of Urban Water Management, Technische Universität Dresden,
Bergstraße 66, 01069 Dresden, Germany
e-mail: benjamin.wagner@tu-dresden.de

F. Al Janabi
e-mail: firas.aljanabi@tu-dresden.de

B. Helm
e-mail: bjoern.helm@tu-dresden.de

C. Koch
e-mail: christian.koch@tu-dresden.de

H. Li · R. Wang · L. Fuchs
Institute for Technical and Scientific Hydrology,
Engelbosteler Damm 22, 30167 Hannover, Germany
e-mail: h.li@itwh.de

R. Wang
e-mail: r.wang@itwh.de

L. Fuchs
e-mail: L.Fuchs@itwh.de

© Springer Nature Switzerland AG 2019
A. Sachse et al. (eds.), *Chinese Water Systems*, Terrestrial Environmental Sciences,
https://doi.org/10.1007/978-3-319-97568-9_3

In the model region of Chao Lake (Chaohu), the fragile interplay of natural and settlement space is currently pronounced and endangered. The region around the lake with the major cities of Hefei and Chaohu City is one of the fastest growing urban areas in the world. The city of Chaohu draws its drinking water from Chao Lake and its further development is strongly linked to its water quality. The increasing anthropogenic pollution components of the lake have led to a significant deterioration of water and water quality in recent years.

An important prerequisite for a better understanding of the lake - city system is the development of an online monitoring system for water quality in the city and in the lake. This includes the monitoring of urban wastewater and water nets. Therefore the listed project partners in Work Package A (WP-A) work together to: install a water monitoring system and coordinate its operation as well as data management (covered by TU Dresden), capture data, modelling and planning the rain water management and sewer systems (covered by itwh), build a data integration (covered by WISUTEC and AMC), organize on-site support (by Tongji University (Shanghai)). The online monitoring stations were installed at different sites and provide important data for model calibration and validation. Within the "Urban Catchments"-project (UC-Project) WP-A covers the urban part of the system analysis and therefore the development of measures for improvements can be considered as the final result.

3.2 Goals

To design effective approaches to improve water quality in urban water and stormwater management supported by an online monitoring system, decision-makers would understand the physical and ecological processes at work, develop a range of possible management tools that are suitable to the site and its problems. To make efficient use of available resources, managers could adopt the water management tools to ensure that targets will be met at the lowest possible cost, especially in using projects technologies for nature-based rainwater and river water treatment. In designing policies to achieve a better understanding for scientist and decision-makers, the focus of the part will be on micropollutants, sewage network protection planning tools and riverbank infiltration (co-funding activities).

3.3 Data Scarcity - Uncertainty and Challenge

Firas Al Janabi, Björn Helm, Benjamin Wagner, Honghao Li, Lothar Fuchs

As long as the possibility exists to reproduce the natural processes (e.g. infiltration of rainwater to the soil) with hydrodynamic rainfall-runoff models the demand for input data is grown with the complexity of the models. A simple split therefore between pervious and impervious areas becomes more divided into smaller subsurfaces. Each of them with its own characteristics (e.g. roughness for overland flow, depression storage). Even if more and more techniques for measuring these parameters are explored, the database for all parameters actually does not exist. So measurements in the field are still needed, that the signals weather from radar, laser and so on could be evaluated and the quality of the data could be secured within a small range of uncertainty. The quality of the model itself has a strong relationship with the data quality. With regard to hydrodynamic simulation of drainage system, lack of pipe network information obviously leads to incorrect results. There is no current Automated Property Map, so real imperviousness could not be determined. Vegetation data gives the basic knowledge of wetting losses and runoff parameters. Without geological data, it is difficult to assess the permeability of the soil. If the data is not sufficient for the simulation, some realistic assumptions will be taken in order to fulfil the requirements. In this case simulation results have uncertainties and the calibration process needs to be carried out.

3.3.1 Drainage System Hydrodynamic Simulation

3.3.1.1 Sewer System

Due to the security rules of administrative authority of Chaohu city, it is impossible to get the current sewer system data for this project. Under this circumstance, the channelization database from 2010 was taken from the previous Chaohu Masterplan Study project, which was carried out by "Wasser Hannover". By analysing the raw database of 2010, some typical conditions regarding to Channelization are listed below:

- Area without sewer (illustrated by Fig. 3.1)
- Unclear flow direction (an example is shown in Fig. 3.2)
- Important information of the sewer system such as elevation data are missing
- Current sewer system has conflicts with existing sewer system on the Masterplan 2030 drawings, such as sewer dimension, flow direction, etc.

For area without sewer system, fictitious sewers were generated according to the drainage masterplan 2030. Unclear flow direction was adjusted through overall analysis of corresponding subnetworks. All above works help to set up a reasonable hydrodynamic model. Blocked sewers could cause the change of flow direction and reduce the discharge capacity, as well as the sediments within the sewers (Fig. 3.3). The current situation of the sewers will affect the evaluation dramatically. Therefore the measured and accurate sewer system is needed for the future work.

3.3.1.2 Catchment

From the raw database, following catchment-related problems are also found:

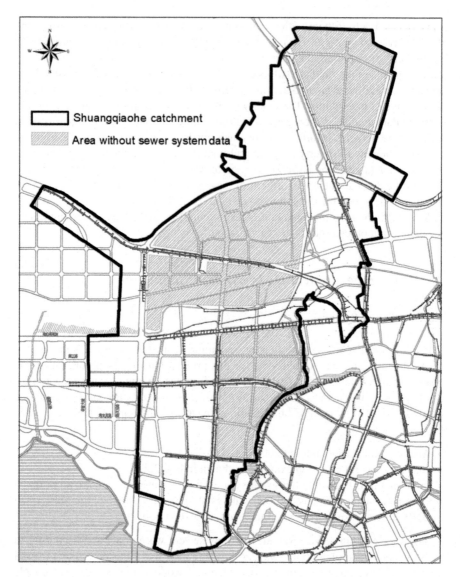

Fig. 3.1 Area without sewer within Shuangqiao River catchment

- No current Automated Property Map
- No current geological data and vegetation data
- Land use conflicts among different drawings

By means of the masterplan 2030 of Chaohu city, Baidu Map, Google Map and the satellite photo, land use types for each area were determined. Referring to corresponding Chinese local technical standards (Municipal Housing and Urban and Rural

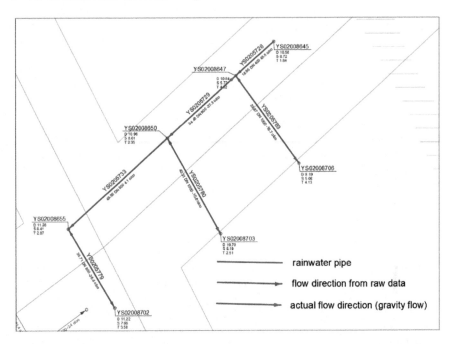

Fig. 3.2 An example on unclear flow direction of sewer system

Fig. 3.3 Example of blocked sewer (**a**) and channelized river. (**b**) with sediments

Planning and Construction Bureau of Zuhai 2015), percentage imperviousness for each land use type was assumed according to the building type and mandatory green area ratio of each land use type. Table 3.1 represents different imperviousness for various land use types.

Table 3.1 Area types and corresponding percentage imperviousness

Land use type	Percentage imperviousness [%]
Bus station, railway station	100
Railway, warehouse, street	90
Industrial/business district, square, parking place	80
Residential district, infrastructure, administrative office	70
Medical use, suburban residential district	60
Unbuilt district (currently green area or uncultivated)	40
Green space	20
Forestry	15

Table 3.2 Standard values of model parameters (Source: itwh.Hystem-extran parameter library)

	Wetting losses (mm)	Depression storage losses (mm)	Initial runoff coefficient	Final runoff coefficient
Impervious area	0.5	1.8	0.25	0.85
Pervious area	2.0	3.0	0	0.6
Street	0.5	0	0.25	0.95

3.3.1.3 Model Parameters

In this project itwh.Hystem-Extran[2] was used for the simulation of sewer system. After building up the hydrodynamic model, model parameters were set in order to run the simulation. Due to lack of geological information of Chaohu city, empirical runoff parameters for each type of areas were selected (Table 3.2), which should be calibrated afterwards.

itwh.Hystem-Extran is the classic option for single or long-term series hydrodynamic simulation of sewer systems. itwh.Hystem-Extran consists of three modules: Hystem-Extran-Editor, Hystem-Extran-Simulation and Hystem-Extran-Viewer, and it supports three simulation methods: ZEBEV, HYSTEM, EXTRAN. itwh.Hystem-Extran stores all incoming data (network data, rainfall data, special profiles, etc.) in a model database and all simulation results in a results database. Results databases from various simulation can be easily compared. After around 30 years continuous running of this hydrodynamic model (itwh.Hystem-Extran) and the improvement of model itself, the simulation results with standard model parameters have been proved to be confident without model calibration.

---

[2]http://www.itwh.de/de/software/software-produkte/produkt-detailansicht/hystem-extran.html.

3.3.2 Data Scarcity for Waterbody Simulation

Thanks to available satellite images it could be shown, how fast Chaohu City is grown and which influence of the Shuangqiaohe riverbed it causes. A residential area should serve as an example as it can be seen in Figs. 3.4 and 3.5. Based on the delineation process for the Shuangqiao River catchment, which is described more detailed in Sect. 3.5, only the riverbed without storage basins and lakes could be done. During project time it was not possible to clarify the water management policy within the storage basins spread in the main city area. Throughout the security guidelines environmental parameters (soil type, evaporation height) are not free available. As a first attempt standard values where chosen for the river model in coordination with the parameters for the sewer system model.

Fig. 3.4 Shuangqiao River bed structure (pink) in comparison to the old satellite images provided by ArcMap Basemap layer based on DigitalGlobe imaginary from 2014

Fig. 3.5 New satellite image provided by google maps (2017 digitalglobe, CNES/airbus)

3.3.3 Outlook

In the further sections, methods and solutions for dealing with the uncertainty and data scarcity problems, are described. With the help of remote sensing data and empirical values, it is shown that, despite a low data basis, modelling is possible, even if the load-bearing capacity and meaningfulness of the model results can only be achieved by means of a reduction with measured data.

3.4 Disaggregation of Rain Data – Methods and Use for Hydrodynamic Modelling

Firas Al Janabi, Björn Helm

The demand for high-resolution precipitation data at temporal scales fluctuating from daily to hourly or even higher resolution is an enormous problem for hydrological modelling. For many locations around the globe, rainfall data quality and quantity are very poor, and consistent measurements are only available at a coarse time resolution. Models for spatially interpolating hourly precipitation data and temporally disaggregating daily precipitation to hourly data have developed for application to multisite scenarios at Chaohu watershed scale. The specialized tool for rainfall disaggregation, in particular at fine time scales, has been examined in more detail. Disaggregation tool called DiMoN based on multiplicative random cascade model used to disaggregate rain data from Chaohu, China whose meteorological data are scarce. A special disaggregation technique, which, instead of using simultaneously both coarser and finer time scales in one mathematical expression, couples independent stochastic model, at each time scale, have been further analysed. According to the absence of hourly data at Chaohu station, data from a station called Luogang, approximately 57 km from Chaohu has been used as a reference station for disaggregation of daily values of Chaohu into hourly and 15 min resolution. Correlation between observed and model-generated data have been found to be 0.84 and 0.77 for hourly and 15 min resolution respectively. NSE (Nash-Sutcliffe Efficiency), RMSE (Root Mean Square Error) and RSR (RMSE-observation standard deviation ratio) show that the model has generated data within an acceptable range. Improvement in the model performance has been demonstrated by the use of finer resolution of longer time series from Chaohu itself.

3.4.1 Daily Precipitation Data

Daily data for six stations (Anqing, Chaohu West, Hefei, Huoshan, Nanjing and Wu Hu) surrounding the study area was obtained from World Meteorological Organisation and National Climate Data Centre, USA (Table 3.3). Figure 3.6 displays a map where all the stations and their approximate distance to the main station (Chaohu West) are shown. Hefei is the closest station while Huoshan is the farthest one from Chaohu. The data quality of the daily precipitation can be classified as a good quality data with only 1.7% of missing values on an average. The next tables contain information about the available values for each station (Table 3.4) and percentage of missing values of each station (Fig. 3.7). Stations Anqing and Hefei have the longest records of available data and the station Wu Hu has the highest percentage of missing

Table 3.3 Status of available dataset

Station	Time series	Missing data [%]
Anqing	1970–2015	1.48
Chaohu west	2000–2015	0.68
Hefei	1970–2015	1.20
Huoshan	1980–2009	1.99
Nanjing	1970–2010	1.94
Wu Hu	2000 (May)–2015	3.77

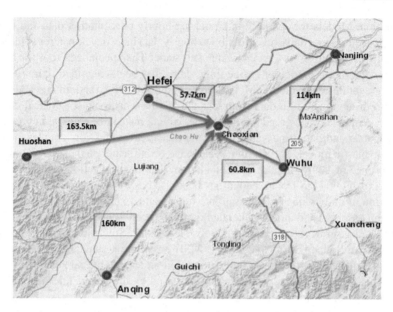

Fig. 3.6 Map showing all the stations and the distance of each station with respect to the main station, Chaohu (Chaoxian) (Source: Google Maps)

Table 3.4 Station details (Source: https://gis.ncdc.noaa.gov/maps/ncei/cdo/daily)

Station	Latitude [°]	Longitude [°]	Elevation [m]
Anqing	30.533	117.050	20
Chaohu west	31.600	117.833	21
Hefei	31.867	117.233	36
Huoshan	31.400	116.333	68
Nanjing	31.933	118.900	15
Wu Hu	31.333	118.350	20

values which is 3.7%. For simplicity in comparison of the stations, the time series from 2000 to 2010 has been taken into account based on the fact that this period has recorded values for all the stations (except Huoshan for which the series is from 2000 to 2009). Within this time frame, all the stations had fewer gaps scattered through the entire time frame.

Figure 3.8 is a graph of daily precipitation of Chaohu West versus time. From the graph, it can be seen that the highest peak of rainfall occurred on 27th August 2008 with a rainfall of 121.92 mm depth, followed by a second highest peak on 23rd July 2008 with a rainfall depth of 105.16 mm. There are seven other peaks with rainfall higher than 60.00 mm depth. Most of the peaks fall in the months of May, June, July, and August with one exception in the month of March (2001) and one in the month of November (2009).

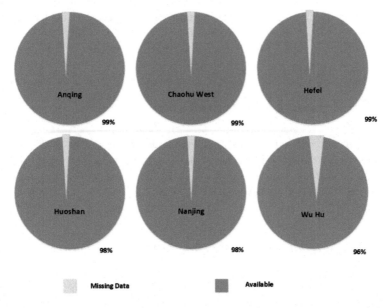

Fig. 3.7 Pie chart showing the percentage of available data from all the stations for the years 2000–2010

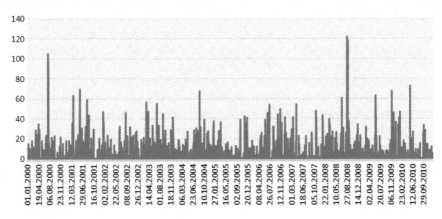

Fig. 3.8 Hyetograph of Chaohu West (2000–2010)

3.4.2 Sub-daily Precipitation Data

Similarly, the hourly reference time series available was the station Luogang. The time series includes precipitation data from 1st January 1984 to 1st January 2014. The data had a lot of gaps and the only period of time that was suitable for use was the time period between 1st March 2010 and 28th February 2011. High-resolution data of 15 min was from the same station from the time period of 1st January 1984 to 1st January 2014. A small part of it was found suitable for further steps which had a continuous data from 1st May 2010 to 2nd September 2010 (four months).

3.4.3 Preprocessing of Rain Data

As discussed earlier, based on the availability of data, the time series 2000–2010 was taken into consideration for this study. This time series includes daily precipitation records of the stations Anqing, Chaohu West, Hefei, Huoshan, Nanjing and Wu Hu (Fig. 3.9). A time period of eleven years has been considered in this study.

3.4.4 Disaggregation of Daily Data to Sub-daily Values

The disaggregation of the daily data from the six stations were disaggregated to hourly scale with the help of DiMoN tool, an application tool for statistical disaggregation of precipitation over the time (Lisniak et al. 2013). DiMoN based on Cascade model of disaggregation which requires reference hourly data and daily data to be disaggregated as input to the model (Fig. 3.10). Firstly, input files were prepared for the DiMoN model. The input files to be used for DiMoN needs to be in a particular format. Ideally, the hourly reference input file needs to belong to the same station whose daily data needs to be disaggregated. However, in this case, due to the unavailability of reference hourly values from Chaohu, hourly and 15 min values from Luogang were taken into consideration. Since DiMoN works on five steps, it was necessary to aggregate the hourly values obtained to 6 hours resolution to disaggregate it into 15 min resolution. Tests were performed to ensure the similarities between Luogang and Chaohu.

Fig. 3.9 Improvement of data quality

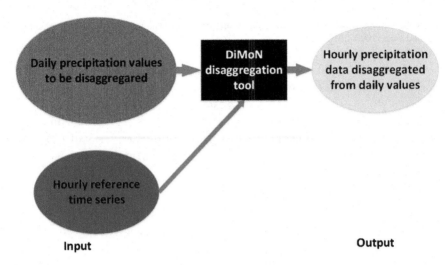

Fig. 3.10 Chart showing the inputs and output of DiMoN model

3.4.5 Disaggregation Output

The daily precipitation from Chaohu West was disaggregated with DiMoN model. The historical reference hourly data was taken from a station named Luogang. The absence of hourly reference from Chaohu West limits the possibility of comparing the disaggregated output and observed hourly data. Therefore, the disaggregated hourly values were aggregated to 24-h duration and then compared with the observed daily precipitation which has been summarized in Table 3.5. It can be seen that the descriptive statistical properties of observed and generated time series are similar in both stations. The difference in the spread (variance) of data is understandable, as the model does not replicate the daily values exactly. Based on the probability of occurrence of a rainy day from the reference hourly data, even a day with rainfall in reality (observed) can get no rain in the generated data series.

In addition to this, the probability of exceeding was also calculated to compare the generated and observed 24-hour precipitation values. The plots in Fig. 3.11 show that the curves were almost replicated by the model for both stations. In case of hourly values, the probability of exceeding of a rainfall intensity of 125 mm/day is 0.02% which is the same in both generated and observed time series. One of the limitations of DiMoN is that it sometimes estimates that the precipitation through the entire day has been collected in one single hour. For instance, the highest rainfall collected in one day for Chaohu is 121.92 mm on 27th August 2008 (Fig. 3.8). On disaggregating, DiMoN calculated that this rain was collected in a time span of 1 hour on the same day between 8 am and 9 am. Therefore, the second peak that occurred on 23rd July 2000 was taken into account for illustration. The observed rainfall depth was 105.16 mm. Figure 3.11 shows how this value was disaggregated through the entire day in which maximum rainfall occurred at 8 am with a precipitation depth of

Table 3.5 Comparison of model generated (hourly and 15 min) and observed 24 h precipitation of Chaohu west (2000–2010)

	Chaohu west (hourly)		Chaohu west (15 min)	
	Observed	Generated	Observed	Generated
Total	10,988.41	10,999.00	10,988.41	10,999.00
Mean	2.7341	2.7368	2.7341	2.7368
Standard deviation	7.8522	8.1075	7.8522	7.9510
Variance	61.6568	65.7317	61.6568	63.2177
Coefficient of variance	2.8719	2.9625	2.8719	2.9053
Skewness	5.6457	5.7523	5.6457	5.7991
Correlation	0.84		0.77	

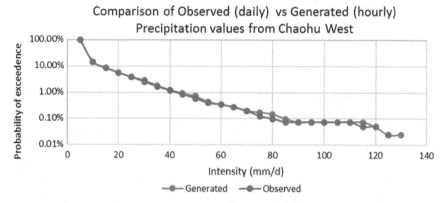

Fig. 3.11 Comparison of observed (daily) and generated (hourly) precipitation values of Chaohu west (2000–2010)

37.3 mm followed by 31.8 mm of rainfall at 10 pm. On adding the total rainfall as generated by the model the daily amount of precipitation amounts to 111.6 mm which is slightly overestimated compared to 105.16 mm observed in reality. Similarly, in case of 15 min resolution, the disaggregated deviates from the observed at a few points.

The NSE for the disaggregation of daily to hourly was found to be 0.65 and for the disaggregation to 15-min resolution was found to be 0.54. NSE equal to 1 means that there is a perfect match between observed and generated values. In this case, the efficiency of hourly time scale is higher than that of the 15 min resolution. This is because the hourly values obtained after disaggregation of daily values have been aggregated to 6 hours time scale in order to obtain 15 min resolution. Therefore, the errors of daily to hourly disaggregation have been propagated to the 15 min disaggregation resulting in a reduced efficiency.

Finer temporal resolution of precipitation data is required in many hydrological studies. The availability of meteorological stations collecting sub-daily precipitation value is scarce compared to the stations maintaining daily values. In such cases, it is possible to disaggregate the coarser resolution data to finer resolution by using reference series from a neighbouring station. The use of DiMoN, a disaggregation model based on the concept of the cascade to disaggregate the daily precipitation values to hourly and 15 min resolution has been demonstrated in this study. In absence of 1 hour and 15 min resolution data from the main station, Chaohu reference data has been used from a neighbouring station, Luogang. A high correlation of 0.84 in the hourly values and 0.77 in the 15 min resolution was obtained. However, in some cases, disaggregation of extreme value was not realistic with DiMoN disaggregating approximately 80% of the rain amount in a one-time unit. The possible reason for this could be the use of a short reference series from a different station. DiMoN can be used as a disaggregation tool and best result can be expected with longer reference series from the station of concern.

3.5 DEM/DTM – Effects on Hydrodynamic Models (e.g. Input Parameter, Runoff and Landuse)

Benjamin Wagner, Björn Helm

The delineation of hydrological catchment areas is one of the most important and fundamental steps in the process of establishing hydrodynamic rainfall runoff models. For the Chaohu City region and the Shungqiao He catchment area, examples of the influence of terrain models with different resolutions and data quality are shown. This also demonstrates the importance of measuring campaigns and data acquisition on site, so that measurement data from remote sensing methods can be validated and the actual goal of identifying the pollutant mass flows can be successfully achieved with the help of the models.

3.5.1 Raw Data

For the model build up usually GIS data for the landuse, sewer system and other typical model parameters provided by companies or the government are used. Time consuming challenges in an international project are at first to build these connections to the regional government. The typical mentality and also the major differences in language make a data handling more complex, additional to the need to manipulate the data to fit in the model structure.

With gathered data from websites or official partners methods of inverse distance weighting (IDW), ordinary kriging, ANUDEM (ANU Fenner School of Environment and Society and Geoscience Australia 2008 and Hutchinson (1988)) and triangulated irregular network (TIN) were used to build digital elevation models (DEM) for the delineation process. The revision process takes place with satellite images from google maps, bing maps and the ArcGIS provided maps.

As a second attempt DEM from radar satellite missions JPL NASA (https://www.jpl.nasa.gov/) with a resolution of 90 × 90 m was used.

During the research project a high resolution DEM could be obtained, so that a 3rd delineation process could take place. With a resolution of 1 × 1 m and the highest trust factor this DEM will be used as the reference, because ground control points were used to check the measured data with a mean range of 0.3 m (Andoczi-Balog 2017).

The links to the hydrodynamic sewer model are the land-use data, maps of the city and the built sewer model as input to avoid doubled surfaces.

3.5.2 Data Manipulation

Based on a study project it showed up, to get DEM out of the elevation contour maps the best results could be obtained by IDW and kriging method. Problems of these methods are so called bulls eyes (e.g. Nusret and Dug (2012)), also the possibility to produce sharp surface intersects like it is common in urban areas. Therefore, satellite images and a visual control of the results are essential.

The evaluation process of the 90 × 90 m ASTER DEM (https://asterweb.jpl.nasa.gov/gdem.asp) was assumed accordingly. General information about the quality of the ASTER DEM could be found in (Tachikawa et al. 2011).

After the setup of the DEM (analogue maps based, satellite or airborne mission based either) the use of hydrological methods as written in Burrough et al. (2015) were done. (e.g. Fill Sink, Flow Directions). The complete process is shown in Fig. 3.12. To end up with a good quality which can be obtained by the method several iterations have to be done. Otherwise, too much computer resources are needed.

As mentioned before (Sect. 3.5.1) additional features for urban areas have to be taken into account:

- sharp change of elevation (e.g. channelized river with rectangle cross section)
- river forced to flow in sewer
- e.g. buildings or bridges causing wrong elevation signals.

So for the urban part several adjustments on the hydrological methods have to be done. The DSM (digital surface model) could be obtained by erasing houses, trees, and so on. Based on the compromise of slope within a certain range some bridges may be not erased in the river. Within the analyses of the depressions the remaining bridges could be identified (Fig. 3.13) and throughout the use of slope filter erased from DSM.

An old but quite important method: visit the area. Pictures of the river, landscape and cross sections are made and taken into account. The use of survey maps is very helpful to check flow directions, estimations for width, and so on.

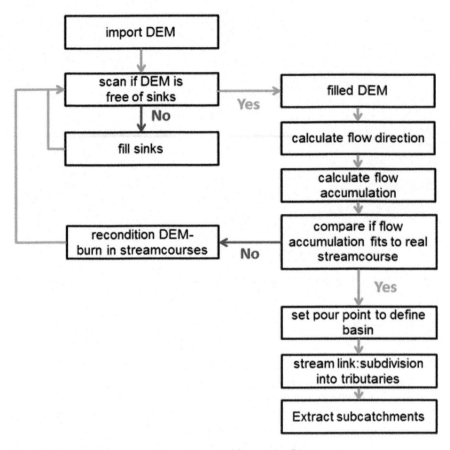

Fig. 3.12 Delineation process of the catchment of Shuangqiao River

To finish the model setup and make it possible to run the model several estimations have to be done. The following list shows the needed parameters and accordingly the source.

- DEM: slope, flow direction, area, subcatchments, cross-section until the water table
- GIS: landuse, impervious area,
- Literature: soil parameters, flow parameters (e.g. roughness factors, depression storage)
- Pictures: cross-section river, soil.

Fig. 3.13 Sinks (green) before filling algorithm

3.5.3 Results

The importance of data evaluation and data plausibility could be seen in the preliminary and final model results (Fig. 3.14).

Even if the calibration process could not be shown as the essential part of the modelling process (Choi and Ball 2002), it is quite important to find out the source of possible pollution. The southern part of the mountain region is more rural and may be an additional source for the eutrophication of the Chao Lake.

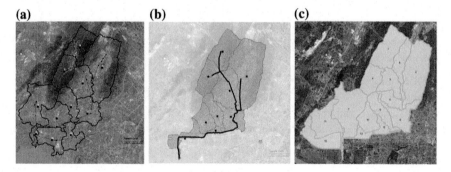

Fig. 3.14 Results of the different delineation approaches: (**a**) IDW, Kriging. (**b**) ASTER DEM from JPL NASA and, (**c**) with high resolution DEM from Airbus

3.6 High Resolution Water Monitoring – A Short Review

Christian Koch, Björn Helm

3.6.1 Introduction

Water monitoring means the collection of chemical, physical and biological parameters within a water body over a certain period of time. In addition, the determination of hydraulic conditions, such as the depth and width of a river as well as flow velocity and flow rate. Primarily, monitoring can be used to assess the current state of a water body. Running over a longer period of time, it can provide information about the development of water quality and is able to prove the success of certain water protection measures that has been taken (Arle et al. 2014).

3.6.2 Types of Water Monitoring

In urban and industrial areas, there are lots of interactions between surface water bodies and urban drainage. Therefore, the observation of quality and quantity of wastewater is also important. Only if both parts are considered and the campaign takes place over several months, a detailed system analysis is possible (Tränckner 2010). In order to detect both continuous and acute loads for the water body, a high resolution is necessary. Typically, water quality measurements are taken at a large temporal distance. The European water framework directive (Commission 2000), hereinafter EU-WFD, specifies three types of water monitoring:

- surveillance monitoring,
- operational monitoring,
- investigative monitoring.

The first one is supposed to be a long term monitoring to evaluate the present conditions and their possible developments. Usually, it is only measured twice in a management period. Operational monitoring takes place at water bodies which are not in a good state. It is done to observe the effectiveness of measures taken to improve the conditions. The resolution is mostly higher than of surveillance monitoring. Investigative monitoring is only done when further information about a water body cannot be obtained with operational monitoring. All these types of monitoring do not have a high resolution, although the EU-WFD recommends the implementation of "more detailed analysis in areas that are protected for drinking water or for natural habitats and species" (Commission 2009).

3.6.3 Necessity of High Resolution Water Monitoring

When trying to understand the linkage between urban drainage systems and surface water bodies such as rivers and lakes, it is necessary to establish a water monitoring with a very high resolution. For example, during a rain event different processes happen simultaneously. To obtain the effects on sewer system and surface waters, it is not sufficient to measure water quality and quantity just once a day or even once per hour. The mostly short-term peaks of flow rate cannot be detected. Thus, information about the amount of an eventual discharge of wastewater into a river and with it the intake of certain loads of pollutants cannot be derived. Another example for the need of a high resolution water monitoring could be the detection of some discontinuous sources of surface water pollution. Whether anthropogenic or natural, it is often not possible to recognize them, when the observation is set up only once a day.

3.6.4 Methods

The two examples mentioned show that continuous high resolution monitoring is necessary, especially in an urban catchment. With this, it is feasible to analyse and compare various chemical, physical and hydrodynamic parameters of different locations and construct a precise water quality model for the area. Sources of pollution or points of unsuitable water flow can be determined and measures to improve the resilience of the whole water and wastewater system can be developed. After their implementation, the monitoring system is able to check the performance of the adjusted network. Thereby, it can help to prove the success and justify investments.

Chaohu City is a highly attractive opportunity to implement a high resolution water monitoring system. As mentioned before, the city is about to change massively. As part of the Anhui Chao Lake Environmental Rehabilitation Project (https://www.adb.org/projects/44036-012/main), additional 100 km of sewer channels will be built. This offers fascinating prospects to run a monitoring system and evaluate the situation before, during and after the project. Especially the impacts on rivers, which discharge into the Chao Lake, seem interesting. It might be shown that fewer pollutants and nutrients enter the lake than before. For this purpose, a monitoring system consisting of four independent but also integrated stations was designed. Two of the stations are equipped for measuring in sewers and two for obtaining information in small rivers. They all provide high resolution data of:

- physical (e.g. temperature and electric conductivity),
- chemical (e.g. concentration of ammonium and COD) and
- hydrodynamic (e.g. water level)

parameters. Those for operation at a sewer are set up with automatic air flush to maximize the required maintenance interval. The two stations, which monitor the

Fig. 3.15 Monitoring station at a small river

river water quality, are additionally equipped with oxygen and pH sensors as well as a flow velocity measurement. One of them is able to observe the phosphorus concentration in a half-hour resolution. The whole system is connected to a web server, which makes it possible to watch live data, control the working modes of the sensors and even to calibrate some of them via remote access (Fig. 3.15).

3.6.5 Outlook

With the high resolution water monitoring system, a "change monitoring" of an urban catchment is possible and thus the control of success can be considered probable. When measured values are compared to other stations, plumes can be traced and flow paths can be detected. Also, an automatically computed mass balance is conceivable. High resolution data is an indispensable requirement for a sufficiently accurate water quality model. Otherwise, there would be no good basis for calibration.

3.7 Drainage System Assessment with Hydrodynamic Model

Honghao Li, Lothar Fuchs

Chaohu city (Fig. 3.16) has been rapidly growing for a couple of decades, and its urban population has increased from 335,830 in 2000 to 404,789 in 2010

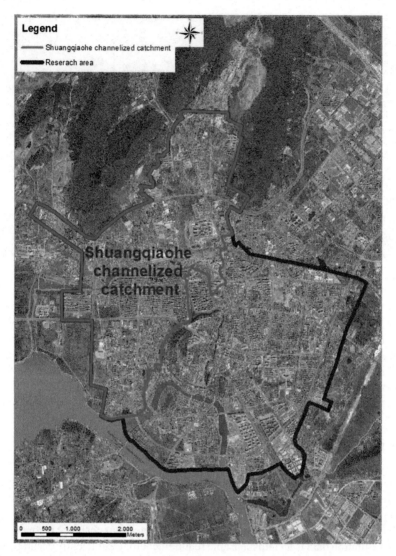

Fig. 3.16 Channelized area of Chaohu city (black) and Study area within Shuangqiao River catchment (blue)

(Population Census data from National Bureau of Statistics of the People's Republic of China[3, 4]). This increase in urbanization leads to a change in land use pattern and an increase urban runoff due to higher proportion of paved area.

---

[3]https://en.wikipedia.org/wiki/Sixth_National_Population_Census_of_the_People%27s_Republic_of_China

[4]http://www.citypopulation.de/China-Anhui.html.

Due to global climate change, the rainfall pattern has been changed and becomes more unpredictable since last 30 years. Therefore previous technical standard used for the urban drainage system design in China has been revised in last years.

Normally the design of drainage pipelines is carried out with the help of the traditional runoff formula, which shows somehow oversized after long term verification. Therefore computational modelling tools are recommended to evaluate the real system and make scientific decisions.

The main goal of setting up the hydrodynamic model of channelization is to assess the real condition of the current sewer system under different circumstances and help to make decisions based on scientific modelling results.

3.7.1 Data Basis

Following data were used to build up the hydrodynamic model in itwh.Hystem-Extran:

- Sewer data
- Area data
- Digital elevation data
- Other inputs data, such as rainfall, inhabitation, etc.

3.7.1.1 Sewer

Sewer data in digital format were collected and transformed into the database. All unimportant branch pipes smaller than DN300 and street inlet junctions were ignored. Overview of the sewer system were illustrated in Fig. 3.17.

Total length of around 200 km was built up in the model, around 50% are the rainwater sewers, 40% are wastewater sewers, and the rest are combined sewers, which mainly located in the old city centre.

3.7.1.2 Area

As there are no topographical maps with detailed estate information, therefore each area with different utilization purpose was regarded as one single catchment. All information of the catchment were stored in the geodatabase, which was created by itwh.FOG[5] based on ArcGIS platform.

Figure 3.18 shows the urban channelization area of Chaohu city and land use in Masterplan 2030. With the help of satellite photos and online maps, such as Baidu Map and Google Map, current land use types for each area were defined.

---

[5]http://www.itwh.de/de/software/software-produkte/produkt-detailansicht/fog.html.

Fig. 3.17 Overview of sewer system in Chaohu city

Different imperviousness (Table 3.6) were determined according to the utilization goal for each type of land use.

3.7.1.3 Digital Elevation Model

To make an analysis of potential flow paths, a digital elevation model in a raster with high resolution is needed. Raster encodes geographic data in the pixel as well as the pixel locations.
Figure 3.19 shows the digital elevation model of Chaohu city. Colour indicates surface elevation value. White represents the highest and black means the lowest.

3 WP-A: Urban Water Resources Management

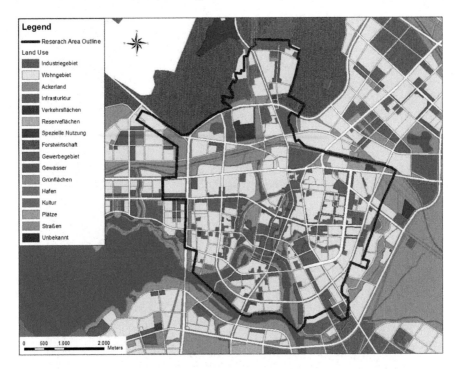

Fig. 3.18 Channelization area (black) and land use of year 2030

Table 3.6 Area types

Land use type	Imperviousness (%)	Area (ha)	Area (%)
Bus and railway station	100	6.87	0.3
Railway	90	9.43	0.4
Warehouse	90	23.80	1.1
Industrial district	80	38.74	1.8
Business district	80	146.07	6.7
Square	80	11.96	0.6
Parking place	80	3.12	0,1
Residential district	70	736.12	33.9
Infrastructure	70	157.17	7.2
Administrative office	70	30.53	1.4
Medical use	60	24.87	1.2
Suburban residential district	60	9.84	0.5
Unbuilt district	40	291.57	13.4
Green space	20	300.56	13.9
Forestry	15	12.84	0.6
Street	90	367.04	16.9
Total		2170.53	100

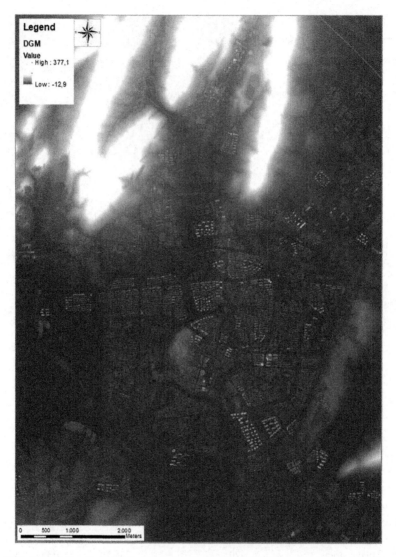

Fig. 3.19 Digital elevation model of Chaohu city (Source: airbus defence and space)

3.7.1.4 Rainfall

As Chaohu city does not have its own rainfall intensity formula, which was generated from long term rainfall statistics, therefore the rainfall intensity formula from the adjacent Hefei city was used to generate the design storm. The distribution of each 5-minute rainfall was created based on Euler type II method (DWA, Deutsche Vereinigung für Wasserwirtschaft (2006); Königer 1981), which is similar to Chicago rainfall pattern used in China (Administration et al. 2014).

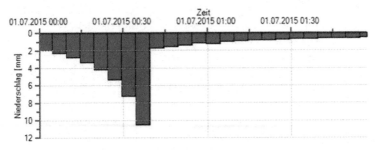

Fig. 3.20 Design storm (return period of 2 a, duration of 120 min)

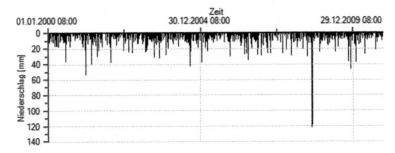

Fig. 3.21 Disaggregated rainfall series for longterm simulation

According to the national technical standard "Code for design of out-door wastewater engineering" (MOHURD 2011) 2016 edition, the return period of 2 years was chosen for the drainage system assessment in Chaohu city.

The design storm with return period of 2 a and duration of 120 min is shown in Fig. 3.20. The generation of model rainfall is based on the stormwater intensity formula from nearby Hefei city. Rainfall peak takes place between 35 and 40 min.

Artificial long term rainfall series were used in order to get the continuous hydrograph of each outlet discharged into Shuangqiao River. The details about the disaggregation of this long term rainfall series can be found in Sect. 3.4.

In Fig. 3.21 the long term rainfall series with duration of 11 years is displayed.

3.7.1.5 Wastewater Information

According to the Masterplan 2030 of wastewater channelization, 5 subcatchments of wastewater within the channelization area of Chaohu city were divided (Fig. 3.22).

Wastewater amount and inhabitants from Wastewater Masterplan 2030 are summarized in Table 3.7. Specific wastewater discharge was about 160–165 l/(cap*d).

Information of 3 industrial point sources were collected and listed in Table 3.8.

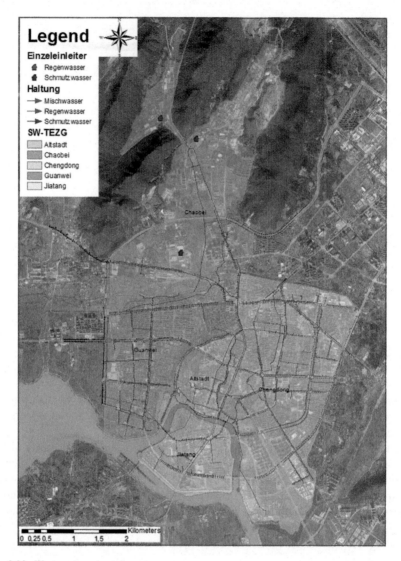

Fig. 3.22 Wastewater subcatchments

3.7.1.6 Model Parameter

As there are no measured geological data within the observation area, therefore the general empirical model parameters were implemented for the simulation. Following parameters in Table 3.9 were used for different area types:

3 WP-A: Urban Water Resources Management

Table 3.7 Wastewater statistics

Nr.	Subcatchment	Area [ha]	Inhabitants	Wastewater from household [10^4 m^3/d]
1	Laocheng	128.2	30000	0.50
2	Chengdong	796.5	144000	2.33
3	jiatang	430.1	82000	1.33
4	Guanwei	194.3	29000	0.47
5	Chaobei	692.3	105000	1.71

Table 3.8 Industrial point sources

Nr.	Company	Amount [10^4 m^3/d]	Amount [10^4 m^3/d]	Discharging sewer
1	Wanwei group	197.6 554.4	2.06	DN 800
2	Chaohu casting group	27.7 13.1	0.11	DN 400
3	7410 Factory	11.7	0.03	DN 500

Table 3.9 Standard values of model parameters (Source: itwh.Hystem-Extran parameter library)

	Wetting losses (mm)	Depression storage losses (mm)	Initial runoff coefficient	Final runoff coefficient
Impervious area	0.5	1.8	0.25	0.85
Pervious area	2.0	3.0	0	0.6
Street	0.5	0	0.25	0.95

3.7.2 Methods

3.7.2.1 Potential Flowing Path

As the geographic data located in the pixel, and each pixel contains one elevation value, potential flow path were determined based on the flow direction weighting method, which is illustrated in Fig. 3.23.

The potential flow paths are essential to determine the natural rainwater watershed. Combined with the analysis of sewer system, realistic catchment for Shuangqiao River was determined afterwards.

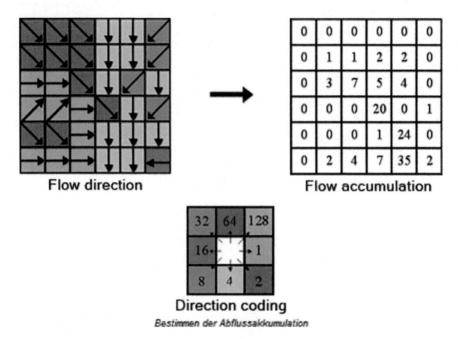

Fig. 3.23 Runoff accumulation according to flow direction weighting (Source: Esri)

Figure 3.24 indicates potential flow paths. Four different colours represent four intervals of grid code value. From green, yellow lines to pink, blue lines, the grid code value continually increase. Bigger value means more runoff accumulation, thus the flow tendency forms.

3.7.2.2 Sewer Topology Check

After importing the sewer data into the geodatabase, topology check was performed to identify the errors in the system. Following conditions were checked within itwh.FOG (Fig. 3.25).

All important errors should be eliminated in order to run simulations. The missing information was complemented by other data source and empirical assumptions. Due to the poor data quality, the longitudinal sections of main sewer were checked to avoid abnormal changes in sewers. In order to get the proper area assignment relationship of the drainage pipelines, some fictitious sewers were generated in areas where information about the sewers was missing (Fig. 3.26). For the newly built sewers, interpolation method was implemented according to the height of the start point and end point.

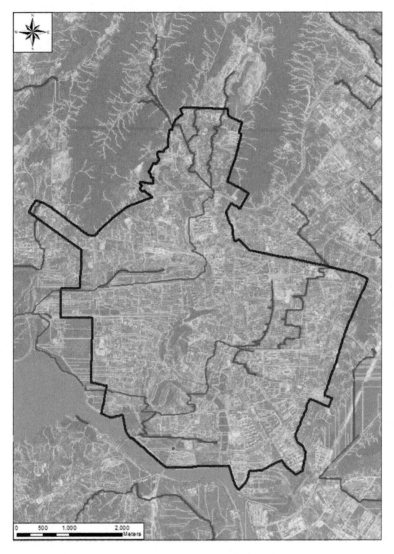

Fig. 3.24 Potential flow paths of Shuangqiao River catchment

3.7.2.3 Area Assignment

As there is no detailed Automated Property Map, which showed the real utilization of the area, accurate area assignment of each estate was impossible. Therefore the general area assignment method of Thiessen Polygons within the channelization area was implemented. After importing the area data into the geodatabase, area topology check was performed in order to avoid area overlaps and gaps between areas. Some green areas adjacent to water body were not assigned to the sewers, and some areas far

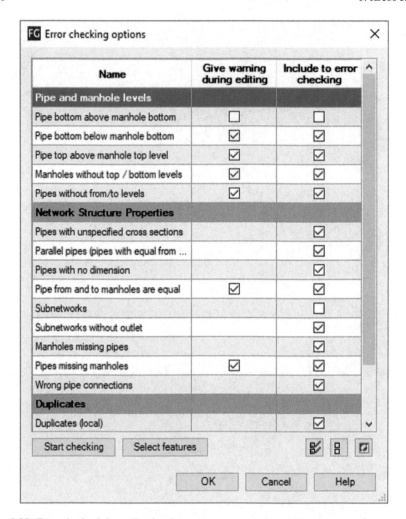

Fig. 3.25 Error check of channelization data

away from the main sewers were not assigned as well. Area assignments were carried out automatically, some unrealistic assignments were manually modified (Fig. 3.27).

3.7.2.4 Dry Weather Flow

Specific water demand of 150l/(cap*d) and 10% sewer infiltration were regarded as input value for wastewater production. All inhabitants were assigned automatically to wastewater and combined sewers except the sewers with length shorter than 5 m. The total inhabitants of each sewer were calculated based on the assigned area and the population density.

3 WP-A: Urban Water Resources Management

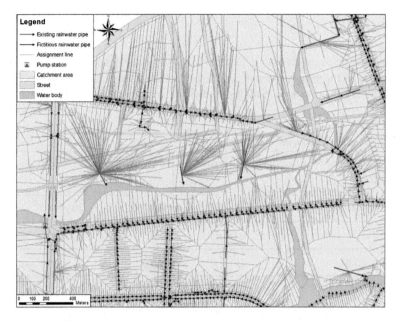

Fig. 3.26 Example of fictitious sewer network

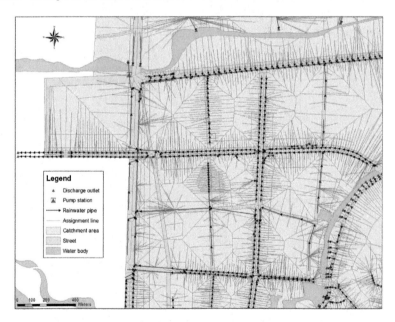

Fig. 3.27 Catchment assignment by Thiessen Polygon for rainwater pipelines

3.7.3 Results

After setting up the hydrodynamic model in itwh.Hystem-Extran, the simulation was executed with the design storm of 2a and the rain duration of 120 min (Abbreviation: T2D120). For the manholes with less than 50 m³ overflow volume, which may cause

Table 3.10 Damage potential classification

Overflow Volume [m³]	Damage Potential	Index	Number of Manholes
<500	Small	I	405
500 – 1000	Middle	II	62
>1000	Severe	III	30

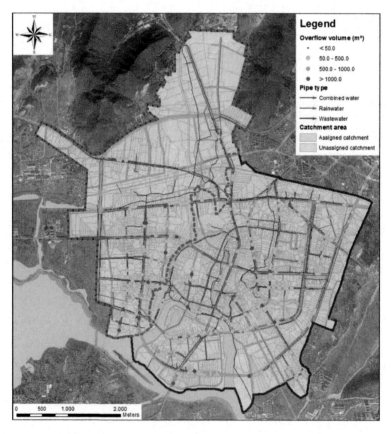

Fig. 3.28 Simulation result of manholes with overflow (T2D120 model rainfall)

really little damage, is regarded as safe manholes. When over-flow volume exceeds $50\,m^3$, it may cause damage to the urban infrastructure. In this project, the damage caused by overflow from manholes were classified into three categories, which are small damage, middle damage und severe damage (Table 3.10).

The hydrodynamic simulation results regarding manhole overflow are shown in Fig. 3.28. Manholes with more than 50 cubic meters overflow are marked in green, yellow and red. Each colour corresponds an interval of overflow value.

As the northern part of Shuangqiao River catchment has the natural slope above 1%, which make the rainwater easier flowing into Shuangqiao River, therefore the overflowing manholes are rare to find. In the southern part of Shuangqiao River, which is really flat compared to the northern part. Under rainfall conditions, the rainwater was pumped to Shuangqiao River. Due to the flow capacity of the pipelines and too many paved areas, large amount of overflowing manholes could be found. Figure 3.29 shows the flow capacity of the sewer system, negative value represents the flow direction is adverse to the sewer's slope. In the figure, green represents no

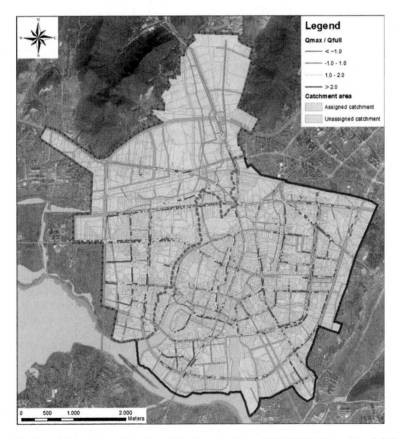

Fig. 3.29 Simulation results of pipeline utilisation rate (Qmax/Qfull, T2D120 model rainfall)

limited gravitational flow in the pipelines, the rest colours represent the pressurized flow. When the flow capacity is strictly limited like the throttle (red lines in Fig. 3.29), which will cause strong backwater effect and overflowing manholes.

In general, the simulation results show the real condition of the current sewer system. With this results, necessary measures should be developed and implemented in order to minimize the damage caused by urban flooding. Further concept regarding to water resource management and improvement of current sewer system should be deepened in order to improve the water environment in Chaohu.

References

Municipal Housing and Urban and Rural Planning and Construction Bureau of Zuhai. Urban Planning Technical Standards and Guidelines, 2015 edn, June 2015

Lisniak D, Franke J, and Bernhofer C. Circulation pattern based parameterization of a multiplicative random cascade for disaggregation of observed and projected daily rainfall time series. Hydrol. Earth Syst. Sci. **17**(7), 2487–2500 (2013). https://doi.org/10.5194/hess-17-2487-2013

Hutchinson MF. Calculation of hydrologically sound digital elevation models, in *Proceedings of the Third International Symposium on Spatial Data Handling*, vol. 133. (International Geographical Union Columbus, Ohio, 1988)

Andoczi-Balog T. Spatiotriangulation report EL1-TERR1-CHINA-CHAOHU_SO17000906 (2017)

Nusret D, and Dug S. Applying the inverse distance weighting and kriging methods of the spatial interpolation on the mapping the annual precipitation in Bosnia and Herzegovina, in *Proceedings of the 6th International Congress on EnvironmentalModelling and Software - Leipzig, Germany - July 2012*. International Congress on Environmental Modelling and Software, Brigham Young University ScholarsArchive, 2012

Tachikawa T, Kaku M, and Iwasaki A et al. ASTER Global Digital Elevation Model Version 2 - Summary of Validation Results. Technical Report. NASA Jet Propulsion Laboratory (California Institute of Technology, 2011)

Burrough PA, McDonnell RA, and Lloyd CD. *Principles of Geographical Information Systems*. Oxford University Press, 2015

Choi K-S, and Ball JE. Parameter estimation for urban runoff modelling. Urban Water **4**(1), 31–41, (2002)

Arle J, Mohaupt V, and Irmer U. Monitoring von Oberflächengewässern in Europa. Korrespondenz Wasserwirtschaft **7**, 379–386 (2014). https://doi.org/10.3243/kwe2014.07.001

Tränckner J. A systematic approach for model-based integrated water resources management, in *Integrated Water Resources Management Karlsruhe 2010: International Conference, 24–25 November 2010; Conference Proceedings*, page 100. (KIT Scientific Publishing, 2010)

European Commission. Directive 2000/60/EC of the European parliament and of the council establishing a framework for the community action in the field of water policy, 2000

European Commission. Water Note 6 - Monitoring programmes. taking the pulse of Europe's waters, 2009

Hydraulische Bemessung und Nachweis von Entwässerungssystemen, 2006

Königer W. Die Anwendung der Extremal-3-Verteilung bei der Regenauswertung und der Niedrigwasseranalyse. gwf-wasser/abwasser, 122 (1981)

China Meteorological Administration and Ministry of Housing and Urban-Rural Development. Technical guidelines for establishment of intensity-duration-frequency curve and design rainstorm profile. http://www.mohurd.gov.cn/wjfb/201405/W020140519104225.pdf, Apr. 2014. In chinese

MOHURD A. Code for design of outdoor wastewater engineering (GB 50014-2006). Ministry of Housing and Urban-Rural Development, General Administration of Quality Supervision, Inspection and Quarantine of the People's Republic of China, 2011

Chapter 4
WP-B: Development and Testing of a GIS-Based Planning Tool for Creating Decentralized Sanitation Scenarios

Thomas Aubron, Manfred van Afferden, Ganbaatar Khurelbaatar and Roland Müller

4.1 Introduction

In Chinese peri-urban and rural areas with low population density, the implementation of central wastewater infrastructure and management concepts is difficult, mainly because of high investment costs and inadequate operation and maintenance concepts. However, these areas strongly contribute to precarious environmental situations of which Chao Lake is a perfect illustration. In the lake catchment three primary diffuse sources of pollution have been identified: (1) indirect discharges of untreated or inadequately treated domestic wastewater, (2) pollution from livestock production, (3) agricultural runoff and groundwater passage. In 2011, these primary nonpoint sources caused 42% of organic (COD) inflows to the lake, 38% of TN, and 42% of TP (Asian Development Bank, ADB 2015). To improve water quality in Chao Lake, investment in wastewater management is likely to be the easiest action as it also helps to improve the quality of the local inhabitants. However, defining and implementing cost-efficient investment plans is difficult and requires enlightened decisions.

T. Aubron (✉) · M. van Afferden · G. Khurelbaatar · R. Müller
Decentralized Wastewater Treatment and Reuse,
Helmholtz Centre of Environmental Research–UFZ,
Centre for Environmental Biotechnology,
Permoserstraße 15, 04318 Leipzig, Germany
e-mail: thomas.aubron@ufz.de

M. van Afferden
e-mail: manfred.afferden@ufz.de

G. Khurelbaatar
e-mail: ganbaatar.khurelbaatar@ufz.de

R. Müller
e-mail: roland.mueller@ufz.de

4.2 The ALLOWS Tool: A Tool for Planning Decentralised Wastewater Management

A key challenge to reduce surface and/or ground water pollutions is to limit or stop the release in the environment of untreated/partially treated or insufficiently treated wastewater. In most situations, the inability to protect the environment from wastewater pollution does not rely on a single issue but rather on the combination of several factors. The main factors to be considered to successfully run a project are:

- **The planning of the infrastructure requirements**: Deciding what needs to be done, in which order (prioritisation) and what the future requirements will be is not easy when dealing with infrastructure whose lifespan can range up to 80 years (sewer network). Planning needs to be methodically carried out as the consequences of poor planning can be felt during several generations.
- **The financing of the infrastructure**: Evaluating the financial burden associated with a specific scheme or technical solution is important to take enlighted and appropriate decisions. This requires an assessment of the capital costs and, sometimes more importantly, of the costs of operation and maintenance activities that have to be carried throughout the entire lifespan of the project.
- **The regulatory and compliance aspect**: Laws and regulations have to be respected, especially when it comes to meeting treatment performance (discharge limits) and obtaining permits and approvals. Monitoring the performance of installed systems is also essential as well as non-compliance protocols to remediate to problems.
- **The technological choices**: Hundreds of technologies are available on the market for collecting and treating wastewater but selecting the best fit to a given situation is not always easy. Collection efficiency and treatment performance have to meet the local constraints such as operators' skills, robustness requirements, maintenance complexity, etc.

All the potential issues related to wastewater management increase in complexity when suburban and rural areas have to be serviced because technical and economic constraints make centralised wastewater management (all-sewer and a single wastewater treatment plant) an unlikely solution. In these situations, stakeholders are often left with the responsibility to improve sanitation without having the tools to do so and without the resources needed to take cost-efficient decisions that will bring the highest benefits (social, environmental and economic) to a region and its inhabitants. To overcome these issues, the UFZ has developed a specific tool called "Assessment of Local Lowest-Cost Wastewater Solutions" (ALLOWS) (van Afferden et al. 2015). ALLOWS enables the compilation of various types of data (see Fig. 4.1) in a Geographical Information System (GIS). The objective is, at a given scale (from local to regional), to create and evaluate wastewater management scenarios that will be compared on a cost-efficiency basis in order to point stakeholders towards appropriate wastewater management solutions. With this approach, the stakeholders are able to refine their scenarios according to local specificities and constraints in order to

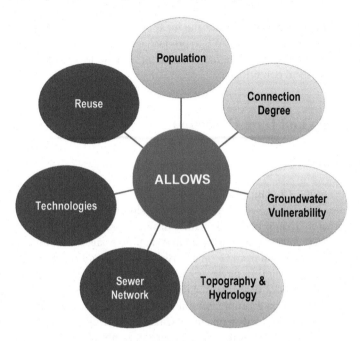

Fig. 4.1 Some of the data fed to the ALLOWS tool

evaluate the financial requirements to meet the objectives (population covered, protection of a specific water body, etc.). This tool also enables the long term planning of the objectives according to economic constraints.

Using GIS enables the visualisation of the information, highlighting particular areas where actions should be prioritised for reasons such as population density, existing health or environmental risks, pollution or contamination of water bodies, etc. For the area of application, the addition of infrastructure data (buildings location, roads, water and sewer networks, existing wastewater treatment plant, etc.) to topography, cadastre and land use data enables the creation of wastewater management scenarios. According to the scale of application (local or regional) and the data available, it becomes possible to identify the buildings and/or villages that can be connected to a single wastewater treatment unit. Scenarios are thus created by changing the level of connection, resulting in shorter or longer sewer networks and in wastewater treatment plant of bigger of smaller capacity. Villages can also be connected in clusters to limit the number of wastewater treatment plants required but at the cost of increasing the length of sewer line and/or the number of pumping stations. The different scenarios created can be evaluated on an economic basis over the project lifespan and the most cost-efficient solutions identified.

Finally, the ALLOWS tool can also help the stakeholders to prepare management schemes for operation, maintenance and monitoring of the planned wastewater infrastructure. Decentralised wastewater infrastructure can then be locally or centrally planned and managed to meet the local conditions (administrative and legal framework). However, for ALLOWS to produce its optimum results, a large amount of data is required and this data has to be as up-to-date as possible to reflect the evolution of the local conditions.

4.3 Surveys and Data Collection

Regional survey

The region around Chaohu has experienced a severe urbanisation during the last couple of decades. This has resulted in a heavy modification of both the city landscape and its perimeter. To apprehend the reality of the situation, a preliminary survey of the area has been carried out using available satellite images and more specifically the historical imagery tool from Google Earth™. This tool allows going back in time and to observe the evolution of the urbanisation process according to the satellite images available. This varies according to the area selected but also according to the scale of the image. As an example, satellite images are available for the entire study region since 1984 but zoomed-in images of Chaohu city itself are only available since 2007. The lower definition of the older images does not allow for a close analysis even though the urbanisation pattern remains visible (Fig. 4.2).

Figure 4.2 shows that Chaohu City has greatly expended but that this expansion has been mostly contained by the northern mountain range. This also shows that thirty years of intense urban development has had a visible impact on the land which will impact the local hydrology and how the pollution concentrates and moves.

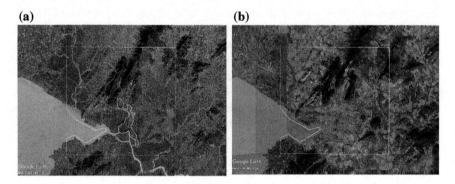

Fig. 4.2 Historical satellite images of the Chaohu region in 1984 (**a**) and 2017 (**b**) from Google Earth™

Site visits

This survey of the historical satellite images allowed the identification of two areas representative of both Chaohu City itself and the rural/sub-urban areas around Chaohu City. These two areas are:

- An urban area close to the heart of the city where new buildings have been constructed near older ones (see red box on Fig. 4.3).
- A rural/sub-urban area located near the lake shore within the economic influence zone (20 km) of Chaohu City (see blue box on Fig. 4.3).

These two areas were then the object of specific site visits to evaluate in more details the local situation. The places visited are marked by red stars on Fig. 4.3. Beyond realising first-hand observations, these site visits were aimed at getting a "feel" for the local situation and, rather than collecting hard data, to engage discussions with the local residents. Specific attention was given to the following elements:

- The type of area (business, industry, agriculture, residential, etc.)?
- The type of population (mostly young, middle age or old people?)
- The social structure (lower, middle or upper class)?
- The condition of the infrastructure (access roads, energy, etc.)?
- The water management (potable water, sanitation, rain water, etc.)?
- The waste management (collection, incineration, etc.)?

Fig. 4.3 Location of the urban (in red) and rural (in blue) area of interest

Discussions with the residents aimed also at understanding what the dynamics of the local areas are in terms of urbanisation and changes in way-of-life. This helped getting a clearer picture of existing local initiatives and if these were meeting the people's expectations while contributing to improve the environment. The discussions finally enabled the identification of some of the administrations and/or institutions involved in the local initiatives discussed and to evaluate if these institutions could be contacted to obtain specific details. This task had previously proved to be challenging from Germany.

Literature and local experience

A difficulty of the project was indeed to access existing hard data about the administrative layout (cadastre, etc.), water quality, water consumption and pollution sources. Although a few publications on the subject exist in English, most of the literature is in Chinese and not necessarily available to foreigners. Of the literature available, several documents were citing governmental or university studies in Chinese, making it impossible to use the mentioned data because of the language barrier, when it was available from outside China. Consequently, contacts and discussions with the local partners and residents were essential to access data and obtain insight on the local situation and policies. Among the key sources of information were partners and contacts from:

- Local administrations and authorities (Chao Lake authority, Chaohu municipality, etc.),
- National and regional Universities (Shanghai),
- Public research centres (Nanjing).

Satellite data

Some of the key information needed to plan decentralised wastewater management are topography-related. Topography defines where the water flows and thus enables the identification of where wastewater can cost-efficiently (limited use of energy or moving-parts equipment required) be transported by relying on gravity. At the regional scale, topography is most efficiently analysed using a Digital Elevation Model (DEM) processed with a GIS. To obtain a DEM without organising a survey expedition, several options exist. The first one is to browse open access (free) satellite data. This option has, typically, a vertical accuracy of 30 m for a horizontal accuracy of a few meters and does not enable the creation of a useful DEM for our purpose. The second option, which we selected, was to purchase satellite data with high accuracy. In this case, we used data from Airbus Defence and Space. The data was bought for an area covering 200 km^2, including most of Chaohu City and the suburban/rural area surrounding Chaohu City (see white frame on Fig. 4.2) so other colleagues from the "Urban Catchments"-project could make use of it.

This data was bought specifically for its one meter (1 m) vertical and horizontal accuracy enabling the creation of a very accurate DEM. This, in turn, allowed the delineation of catchments and the identification of water runoff flow paths which led to the localisation of potential pollutants flow path and/or pollutant sources.

4.4 Scenario Preparation

Once the topographical data had been collected, it had to be aggregated and processed in order to prepare the wastewater management scenarios. While the site visits helped to figure out which technical solution could be feasible, it was only after the data processing that the management scenarios were defined. Typically, tens or hundreds of different scenarios can be created from the data set by changing the potential clustering of houses and villages. However, framing conditions should be used to reduce the scenario number to a manageable amount, enabling meaningful comparison. Usually, between three (3) and five (5) scenarios are created.

Data aggregation

Data aggregation consists in combining all the data available in GIS. Table 4.1 lists two types of data: the data necessary (mandatory) and the data whose availability is not critical but highly beneficial to a project. The table finally shows the data availability for this project.

As can be seen in Table 4.1, it has not been possible to obtain the entire dataset for the "Urban Catchments"-project with the consequence that other sources of information had to be found (i.e. satellite images to identify roads and buildings) and that assumptions had to be taken. The assumptions taken to overcome the missing/partially available data are discussed in the presentation of the scenarios.

Data processing

Once the data was aggregated, it had to be processed (Table 4.2). The first processing steps define the location of houses and villages within the landscape (topography and hydrology) while further processing steps deal with the organisation/connection of the buildings between themselves (clustering) and the definition of technological specification.

The design of potential sewer network is critical to identify if a specific building and/or a specific village can be connected to a sewer network that will bring the produced wastewater to a single location where a WasteWater Treatment Plant (WWTP)

Table 4.1 List of data to be aggregated for the creation of wastewater management scenarios

Type of data	Requirement	Project availability
Digital elevation model (DEM)	Mandatory	Yes
Buildings (DEM)	Mandatory	Incomplete
Road network	Mandatory	Incomplete
Protection zones/regulations	Optional	No
Cadastre and planning	Optional	No
Administrative boundaries	Optional	No
Sewer networks and wastewater treatment plants	Optional	Incomplete
Land use	Optional	No

Table 4.2 Data processing steps and type of action

Action order	Type of action
Defining settlement boundary	Data processing
Computing micro-catchments	Data processing
Computing building density	Data processing
Identification of building with potential connection to sewer networks	Network design
Identification of potential location for wastewater treatment plants	Network design
Selection of buildings and/or villages clustering	Scenario definition
Selection of size and location of wastewater treatment plants	Scenario definition

can be built. This potential buildings connectivity enables the simulation of different sewer options and permits to modify the location of WWTPs. These simulations are compared and the realistic options are turned into scenarios. In the absence of cadastre, WWTP locations were selected as unbuilt area within or at a reasonable distance of villages.

Scenario analysis

The scenarios defined through the data processing step can then be analysed and compared. Evaluation is made on a technical level (e.g. length of sewer line, number of pumping station, number and size of wastewater treatment plants, type of technology for the treatment of wastewater, etc.) to determine the components required for each scenario. A cost assessment of each scenario can then be performed and the results compared to identify the best option(s). Recommendations can finally be made on the best option to manage wastewater at the selected scale (village, region …). The preferred options are not always the cheapest to build as operation and maintenance complexity and costs over the life span of the system are considered.

4.5 Wastewater Management Scenarios for the Test Areas

4.5.1 Scenarios for the Urban Areas

Urban context

The first set of scenarios focuses on the urban area where it has been observed that old villages had been swallowed by urban extension. During site visits and interviews with inhabitants, it has been found that these villages had not been connected to the sewer installed for the new developments/suburbs built nearby and, as such, wastewater management in these villages remains an issue. Typically, in these urban villages, open pit-latrines are used for faeces with the leachate infiltrating into the soil and potentially contaminating groundwater and/or nearby waterways. When full, these latrines are emptied by nearby farmers that use the "compost" as crop fertilisers. The remaining wastewater (grey water) produced by the households is discarded in rainwater drains or sent directly into nearby waterways. From the satellite data analysis, it is estimated that at least 13,000 people are living in these urban villages. This number was calculated by multiplying the number of buildings identified in the delineated urban villages by the number of inhabitants per household, using the assumption that one building equal one household. These 13,000 people live in an area of $1.2\,km^2$ equal to a density of 10,800 habitants per square kilometre. This high density support considering these villages as a source of pollution of the urban waterways and justifies the use of ALLOWS to test wastewater management scenarios in urban/peri-urban areas. The visits of these urban villages also highlighted the fact that these villages cannot be considered as industrial or commercial centres. It is thus assumed that the wastewater found in these villages can be classified as domestic wastewater.

Urban topography and hydrology

Figure 4.4 shows the DEM that was created for the Chaohu area from the 1 m-accuracy satellite data. The DEM does not consider infrastructure such as buildings, bridges, roads and railways in order to avoid interference with the hydrology of the region. However, the berms and/or ground elevations on which the infrastructure is built are correctly represented.

This DEM shows, as expected, a gradient from the slope that goes from the mountain range to the lake. It also shows that the area bordering the right bank of the Shuangqiao River is quite flat. It is the location of most of the "urban villages" identified during the site survey. On the downstream part of the river's left bank, a berm following the river modifies the topography of the land. The hydrological analysis of the urban area shows that the "urban villages" previously identified are all part of the Shuangqiao River catchment. Several micro-catchments have been computed by analysing flow direction and flow accumulation, resulting on "topographical streams" that bring runoff water to the river. Overall, the computed streams are close to the streams observed during site visits or on satellite images.

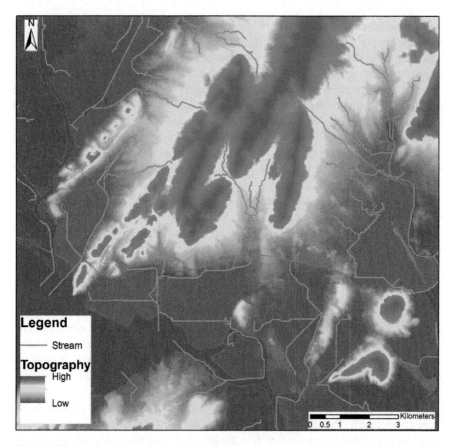

Fig. 4.4 DEM and hydrological map of the Chaohu area (© Cnes 2017, Distribution Airbus DS)

Figure 4.5 shows the Shuangqiao River catchment and its subdivision into micro-catchments in the different colour shades on the map. These micro-catchments are those contributing to the discharge points observed where wastewater is released into the ShangQiao River. Some low-lying areas are not included in the micro-catchments as the computation estimated that these areas are not contributing to a discharge point but instead to the ShangQiao River itself.

As can be seen on the map, the localisation of the villages in the river catchment gives four potential situations:

- A village is included in its entirety in a micro-catchment.
- A village is split between two or more micro-catchments.
- A village a village is located in the river contribution area.
- A village is split between one or several micro-catchments and the river contribution area.

4 WP-B: Development and Testing of a GIS-Based Planning ...

To simplify the scenario construction, village unity has be conserved and a village split between one or more micro-catchment has been allocated in its entirety to the catchment where most of the village is located. Three scenarios have been devised. They are presented in Sect. 4.5.2–4.5.2 along with the rationale behind their creation and the specific decisions for the grouping of villages.

4.5.2 Urban Wastewater Management Scenarios

Urban scenario 1: Low clusterisation

Figure 4.6 shows a map of the first urban scenario. In this first scenario, the villages are grouped in clusters if they are geographically close and if they are not separated by existing infrastructure such as highways or railroads. The use of pumping stations and pressurised sewers have been implemented but reduced to the minimum through the use of gravity. The WWTPs have been located outside the villages on non-built area

Fig. 4.5 Hydrological map of the urban area

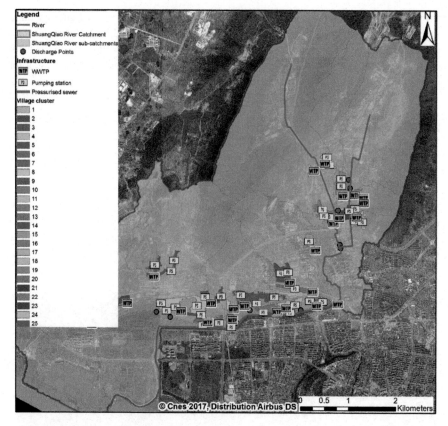

Fig. 4.6 Map of urban scenario 1

and when possible, at the lowest location. WWTPs have also been located as close as possible to the wastewater discharge points that had been previously observed.

Urban Scenario 2: High clusterisation

Figure 4.7 shows the map of the second scenario where villages have been clustered to a higher degree. In this scenario, villages that are geographically further apart have been connected together but not if they were separated by existing infrastructure (highways, railroads …). This scenario has the interest of reducing the number of WWTPs but the adverse effect is an increase in the number of pumping stations and the length of pressurised sewers to connect the villages together and to the WWTPs. Here again, individual villages are connected to a single WWTP. The WWTPs are also located at the lowest location near the discharge points to make the best use of gravity.

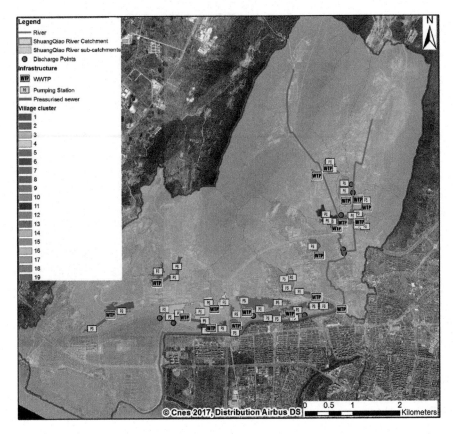

Fig. 4.7 Map of urban scenario 2

Urban Scenario 3: Clusterisation and connection to municipal sewer

Figure 4.8 presents the third urban scenario. This scenario is similar to the Scenario 2 except for the villages adjacent to a main road or a highway connecting newer part of the city. In these situations, it is expected that municipal sewer mains are located along the roads and highways and that the villages can be connected to these sewer mains. The result is a steep reduction in the number of WWTPs but a somehow similar number of pumping stations. The municipal sewer mains and the municipal WWTP are expected to be able to handle the additional wastewater inflow without requiring upgrades. For the villages not connected to the sewer mains, the location and connections to the WWTPs follow the same rules as in Scenario 2.

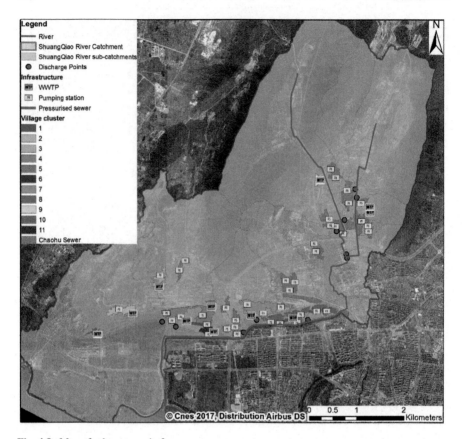

Fig. 4.8 Map of urban scenario 3

4.5.3 Results for the Urban Scenarios

For all the urban scenarios, the connection of the buildings to the closest sewer lines have not been technically or economically considered as too many parameter related to the specific building design enter into consideration. Additionally, the WWTPs capacity (in Person Equivalent (PE)) has been oversized by 20% to acknowledge the potential growth of the connected population and the fact that some economical activities (restaurants, shops, etc.) may occur and result in an increase of wastewater production. This 20% increase in plant capacity serves also as a safety buffer for the design and operation of the WWTPs.

Components analysis

Table 4.3 presents the different technical elements that would need to be built for each urban scenario. These elements are the number of WWTPs, split in different size

Table 4.3 Description of the main components of the urban scenarios

Item	Item	Urban scenario 1	Urban scenario 2	Urban scenario 3
Number of WWTP	$X < 10$ PE	1	1	1
	$10 \leq X < 50$ PE	1	0	1
	$50 \leq X < 500$ PE	7	5	4
	$500 \leq X < 2000$ PE	16	13	5
	$X \geq 2000$ PE	0	0	0[a]
	Total	25	19	11
Length of pressurised sewer (m)		1990	2680	1200
Length of gravity flow sewer (m)		57,800	59,400	58,900
Number of pumping stations		37	42	41
Number of manholes		1180	1230	1200

[a]This scenario requires the connection of an additional 6600 PE to the Chaohu municipal WWTP. In this scenario, we consider that the existing infrastructure (sewer and WWTP) can handle the additional volume and load of pollutants

categories, the length of gravity and pressurised sewer and the number of pumping stations and manholes.

The first scenario require the construction of 25 WWTPs of which one is considered an onsite system ($X < 10$ PE) and one is a very small decentralised system (between 10 and 50 PE). The other 23 WWTPs are of small or medium size but no system reaches the 2000 PE size. This first scenario also requires the construction of c.a. 2 km of pressurised sewer and 58 km of gravity sewer as well as 1180 manholes. This is high sewer length is acceptable to cover almost 16,000 PE overall. The high number of WWTPs has for consequence to limit the number of pumping stations required to 37.

The second scenario limits the total number of WWTPS to 19 with 70% of the WWTPs being in the range between 500 and 2000 PE. Here as well, no WWTP above 2000 PE has to be built and only one onsite system is required. In this scenario, the number of pumping stations required to connect the villages together and to the WWTPs reaches 42 and the length of sewers c.a. 2.7 km (pressurised) and 60 km (gravity). 1230 manholes would also have to be built.

Finally, the third scenario calls for only 11 WWTPs of which one is an onsite system and 50% of the remaining are of the class 500–2000 PE. It is important to note that in this scenario, half of the population would be connected to the municipal WWTP of Chaohu City and that it is expected that no upgrades would be required for both the WWTP and the sewer lines. This scenario calls for the construction of

Table 4.4 Cost comparison of the urban scenarios

Costs	Station	Urban scenario 1	Urban scenario 2	Urban scenario 3
Capital costs ($)	WWTP	9,110,425	8,894,981	4,428,201
	Pumping stations	575,056	652,766	637,224
	Pressurised sewer	1,309,632	1,194,966	743,116
	Gravity sewer	17,757,447	17,700,259	17,557,289
O&M costs ($) (year^{-1})	WWTP	1,215,727	1,279,223	656,062
	Pumping stations	3016	4157	4360
	Pressurised sewer	7335	6693	4162
	Gravity sewer	208,569	207,869	206,119
Capital costs total ($)		28,752,560	28,442,973	23,365,830
O&M costs total ($)		31,053,407	32,423,453	18,846,644
Net present value ($)		59,805,968	60,866,426	42,212,474

41 pumping stations, 1.2 km of pressurised sewer lines, c.a. 59 km of gravity sewer and 1200 manholes.

Table 4.3 shows that the two first scenarios are quite similar. This is due to the decision to not connect villages separated by highways and railroads to avoid major construction challenges. The third scenario is quite different but relies on the assumption that the existing municipal infrastructure can handle the additional wastewater flow. Overall, the third scenario seems less complex to implement on an organisational basis (operating 11 new WWTPs is not as challenging as operating 25) and may be more efficient to limit river/lake pollution by limiting the number of WWTPs and thus the risks of system failure.

Costs analysis

Table 4.4 presents the results of the costs estimations calculated according to the technical specifications presented in Table 4.3. For this analysis, the following assumptions were taken:

- The unit costs for each technological component are from a database compiled by the UFZ through several similar projects in countries such as Germany, Jordan, India, Mongolia and Oman.
- The costs are presented in US dollars for easier comparison.
- A scenario's Net Present Value (NPV) is the total cost of a scenario over the life span of the project brought back at the cost of today. In more details, the NPV is the sum of the capital costs, paid once at the beginning of the project, plus the O&M costs that are paid every year during the lifetime of the project but calculated if the entire sum had to be paid today.

- The discount rate used for the calculations was set at 2.25%, as per data from the central bank of the People Republic of China.
- The life span of the WWTPs was set at 30 years. Reinvestment in WWTP has not been considered after 30 years as it is not clear which technology would be selected or if a connection to the municipal WWTP would be preferred.
- The life span of the sewers was also set at 30 years old. The typical lifespan of a sewer network is usually around 80 years but given the local urbanisation dynamics and the current construction materials used and techniques employed, the sewer network is not expected to last beyond 30 years without heavy upgrades that cannot be considered in this situation.
- The life span of the pumping stations was set at 30 years with a reinvestment every 5 years in moving parts, electrical equipment and water tightness. This reinvestment has been included in capital costs instead of Operation and Maintenance (O&M) costs for facilitating the calculations.
- For the treatment systems below 10 PE, septic tanks and dispersal fields have been selected as the single technology. This decision was made because septic tanks design is quite common and easy to replicate and install. The use of dispersal fields ensures that the treated effluent is spread over a big enough area enabling the remaining nutrients to be used by plants and soil organisms.
- A single wastewater treatment technology was selected for all systems above 10PE: the Sequencing Batch Reactors (SBRs). The reason is that SBRs are space efficient which is of high importance in built-up environments where space is at a premium and usually expensive. SBRs have also the capacity, if operated correctly, to reduce nitrogen by up to 80% by performing both nitrification and denitrification. A drawback to the use of SBRs is that they are energy intensive, relying on pumps and aerators that consume electricity and are also O&M heavy as they require frequent cleaning and change of spare parts.
- Finally, scenario 3 considers that the municipal WWTP of Chaohu City is able to cope with the connection of an additional 6600 PE.

Table 4.4 shows that, as expected, the scenario 3 is the cheapest option. The main reason is because the WWTP costs (capital and O&M) are greatly reduced compared to the scenarios 1 and 2. The costs of scenarios 1 and 2 are quite similar (1.6% difference) but in scenario 2 (high clusterisation), the higher number of pumping stations overcomes a lower cost for WWTPs and makes this scenario slightly more expensive.

Recommendations

Scenario 3 seems the most feasible option on both an economic and technical basis as it limits the number of WWTPs to be built and reduces both the investment in capital and the O&M costs. Scenario 3 also limits the number of WWTPs to be built and thus lower the risks of pollution associated to the potential failure of treatment systems. If for some reasons the scenario 3 becomes unrealistic, the scenario 2 would be the preferred replacement option as it has a lower number of WWTPs to be built (19 instead of 25) than scenario 1, even if it is marginally more expensive. The lower number of WWTPs in scenario 2 would also facilitate both maintenance and monitoring activities and reduce the risk of system failure. The low space available at the building level in the urban environment renders impossible in most of the cases the installation of onsite wastewater treatment and thus forces the collection and transport of raw wastewater to an outside area for treatment. In every scenario, installing a sewer network in urban villages is expensive. However it is necessary to limit the discharge of untreated wastewater into water bodies. A sewer network would also help to improve the sanitary situation in the urban villages where pit latrines have been observed and present a risk to the local inhabitants. Finally, preparing another scenario where the urban villages would have been connected together independently of the existing infrastructure (highways and railroads) would have likely shown an interesting result on both a technical and economic level. However, the feasibility of such a scenario is difficult to assess as closing down roads, highways and railroads always presents significant constraints that local administrations are usually reluctant to experience.

4.5.4 Scenario for the Rural Areas

Rural context

The second set of scenarios focuses on rural/peri urban areas. Some of the villages in these areas a experiencing depopulation as younger people leave to find work in urban areas to come back only a few times a year during holiday time. Some other villages manage to maintain their level of population, mostly employed in various agricultural activities. However, the population of these villages has still expressed the need to improve their current quality of life and especially the local water management. Indeed a common observation was that these villages experience a low or inexistent coverage by wastewater management infrastructure. The sanitation situation in some of these villages is poor with some households still relying on buckets to get rid of night soil. In other cases, the situation is very similar to the urban villages with pit latrines being used and emptied regularly by farmers using the "compost" as crop fertiliser. In these villages the contamination of waterways by domestic wastewater is expected and is likely to exacerbate diffuse pollution by agricultural fertilisers. A specificity of these rural areas is that the local population seems to live exclusively in villages or hamlets ranging from a couple to a few tens of households. The bigger villages may be fitted with a couple of commercial buildings (shops or restaurants) but this remains a rare occurrence. The local wastewater can thus also be classified as domestic wastewater. As such, these rural/peri-urban area are also the perfect situation to use the ALLOWS tool to test wastewater management scenarios.

Rural topography and hydrology

Figure 4.4 previously showed the DEM that was created from the 1m-accuracy satellite data. This DEM also includes the rural area of interest (in the North West corner of the map). Again, this DEM does not consider the existing infrastructure to limit the impact on the hydrology of the region. However, the berms and/or ground fluctuations on which the infrastructure is built are correctly represented. The DEM shows that the topography of the rural area of interest is quite flat. The DEM also shows that the rural area is generally low lying and has an elevation close to those of the nearby river and of the lake. The area of interest shown on Fig. 4.9 is delineated by two rivers coming respectively from the north-east and north-west and connecting at the south of the selected area. A highway marks the northern border of the area. Both rivers have a berm on their bank that isolate the area from the rivers themselves while the highway has been built on pillars and, as such, does not affect the local hydrology. The hydrological analysis of the rural area showed severe inconsistencies with the visual observation (satellite images) of the area:

- The water from the streams and canals flowed towards the north while Chao Lake is located in the south where the two rivers are in fact flowing.
- Most of the computed waterways don't match the observations from the satellite images.

The discrepancies between the computed delineation of the waterways and the visual observations are attributed to the flat topography where a 1 m elevation accuracy is not enough to correctly represent the shape of the land. Only a field survey would be able to identify the topographical specificities. Hence, the DEM was not used for the hydrological analysis. Instead, satellite images were used to delineate the dense network of streams and canals (Fig. 4.9).

The three scenarios prepared are presented in Sects. 4.5.5–4.5.5). The specificities of the scenarios and the rationale behind their creation is discussed in the respective subchapter. However the following topographical and hydrological conclusions were drawn:

- The reduced slope constrains the use of gravity flow sewer.
- The high water table presents a challenge for the installation of below-ground infrastructure (sewer, collection tanks …).
- The high water table presents a risk of either or both ground water infiltration into the sewer or of sewage contamination of the groundwater.

From these conclusions, the following assumptions were made:

- The installation of sewer lines deeper than 1.5 m would present too much risk of pollution or groundwater infiltration and should be avoided.
- The slope of the sewer line should not be lower than 1%. A minimum of 1.5% would be preferable but a 1% slope will be tolerated for length of up to 100 m.
- The installation of sewage pumping station will be more frequent than is common. The water proofing of these pumping stations will have to be carefully done to avoid both ground water infiltration into the sewer or of sewage contamination of the groundwater.

Additionally, without access to the local cadastre, it was not possible to identify the land where wastewater treatment systems could be installed (public land …).

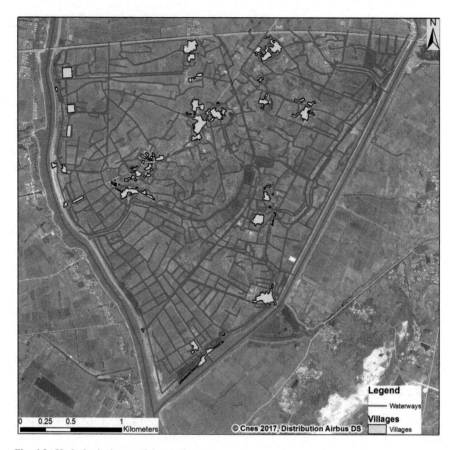

Fig. 4.9 Hydrological map of the rural area

As such, it was assumed that the wastewater treatment systems would be built at the most convenient location (reduced sewer length, close to the villages' centres of gravity, etc.). As a result, a 100 m radius circle was extended around the centre of gravity of each village to identify the maximum reach of gravity flow sewer lines. Three different situations then occurred:

- A specific village was entirely covered by a circle and did not intersect another circle: a gravity flow sewer and a single wastewater treatment plant were considered.
- Two or more circles intersected and covered all the neighbouring villages: a gravity flow sewer would be installed and a single WWTP would be built to service these villages.
- Two or more circles intersected but did not cover all the neighbouring villages: gravity flow sewer would be installed to cover as much of the area as possible. The effluent would be collected in a pumping station and sent through pressurised

sewer to the WWTP. A reduction of the length of pressurised sewer and of the number of WWTPs is always deemed to be the best option.

From these results, three wastewater management scenarios were prepared and are presented below.

4.5.5 Rural Wastewater Management Scenarios

Three scenarios have been prepared for the rural area: high decentralisation, low and high clusterisation. These scenarios purposefully present very different wastewater management possibilities that range from one extreme (onsite wastewater treatment)

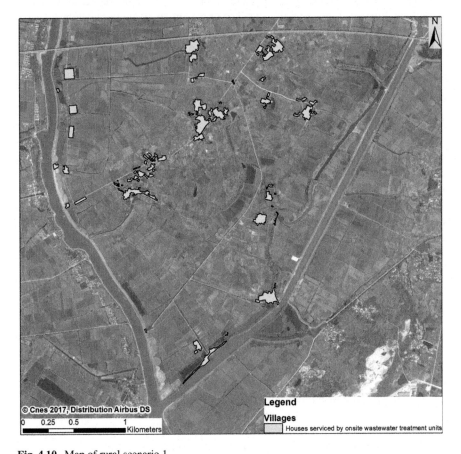

Fig. 4.10 Map of rural scenario 1

to the other (high clusterisation). A scenario presenting the centralisation of wastewater collection and treatment in the study region has been avoided as such scenario would be unlikely to be either technically feasible or affordable.

Rural scenario 1: Onsite wastewater treatment and reuse

The first scenario prepared for the rural area avoids completely the use of pumping stations and wastewater collection infrastructure (sewer) by requiring each building of the villages to be connected to their own wastewater treatment system. Figure 4.10 shows a map of the test area where all the buildings of the villages are connected to their own onsite treatment systems. The use of onsite wastewater management is common in rural areas where space is usually available for installing the treatment technologies and where the treated effluent can be reused in nearby lands (usually for tree irrigation). This approach typically requires robust technologies that usually have a lower treatment performance and relies on the wide area available to spread the remaining nutrients to avoid pollution. The drawback to onsite wastewater management is that the number of WWTPs can be high and that the type of technology used usually produces sludge that needs to be pumped out regularly (every six to twelve months and disposed of in an appropriate facility. In onsite wastewater management schemes, the monitoring and control of the systems is critical to avoid pollution and sanitary risks.

Rural scenario 2: High decentralisation

The second rural scenario presenting high decentralisation is shown on Fig. 4.11. In this scenario, one single WWTP per village is envisioned to treat the wastewater produced. As such, the villages are fitted with sewer lines but these are kept to the minimum. The use of pumping station and pressurised sewers is also avoided as much as possible to limit the technical complexity (construction, O&M, etc.) and to keep the costs as low as possible. Nonetheless, as show on the figure, it is obvious that a high number of WWTPs would have to be built, having an impact on the cost of the scenario but, more importantly, the aspects related to the operation and maintenance of the WWTPs will be complex to handle, increasing potential risks of failure and environmental pollution.

Rural scenario 3: Clusterisation

Finally the third scenario is presented in Fig. 4.12. In this scenario, the villages are grouped together in clusters as much as possible with pressurised sewer lines and pumping stations to keep the number of WWTPs to a minimum. This scenario represents one way to connect the villages together but other clusters would have been possible. The low number of WWTPs is counterbalanced by a high number of pumping stations and longer distances of pressurised sewer. In this scenario, operation and maintenance of the WWTPs would be simplified while the maintenance of the pumping stations and pressurised sewer lines would be more complex.

4 WP-B: Development and Testing of a GIS-Based Planning …

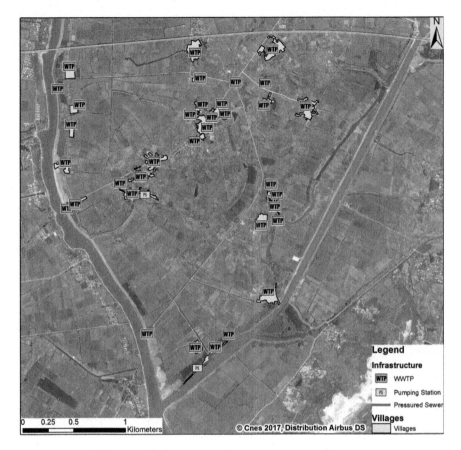

Fig. 4.11 Map of rural scenario 2

4.5.6 Results for the Rural Scenarios

Before detailing the technical and economic evaluations, a couple of points have to be mentioned. The first point is that, as for the urban scenarios, the rural scenarios are not considering the connection from a building to the closest sewer line. The second point concerns the capacity of the WWTPs. This capacity (in PE) has been increased by a 15% factor to cover the potential growth of the villages and to ensure a safety buffer volume during operation.

Components analysis

Table 4.5 presents the technical components of the different scenarios created. These components include the number of WWTPs, the length of gravity and pressurised sewer lines and finally the number of pumping stations and manholes. In more details, the number of WWTPs is split between the same five classes of sizes as for the urban

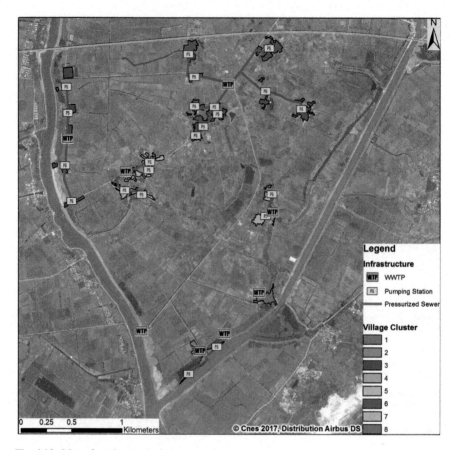

Fig. 4.12 Map of rural scenario 3

scenarios. These classes have an impact on the technical specificities and the costs of the treatment systems.

For the first scenario, 865 WWTPs (onsite systems) would have to be built. No sewer lines, either gravity or pressurised, would be built, avoiding also the need for pumping stations and manholes. For the second scenario, 36 WWTPs would have to be built, the smaller systems servicing 3 PE (onsite systems) and the bigger 290 PE (small decentralised WWTP). The total length of sewer lines to be build reaches 12,200 m (12.2 km) of which 11,700 m (11.7 km) are gravity lines and almost 500 m are pressurised lines. For the sewer network, 250 manholes would have to be built (1 mahnole every 50 m approximately). Finally, two (2) pumping stations would have to be built. Finally, the third scenario calls for only eight (8) WWTPs. Of these, two would be onsite systems, five WWTPs would be of the 50–500 PE range while the last WWTP would be sized for approximately 1430 PE. This scenario calls for the longest sewer network with 5590 m of pressurised sewer as well as 11,620 m of

Table 4.5 Description of the main components of the rural scenarios

Item	Item	Rural scenario 1	Rural scenario 2	Rural scenario 3
Number of WWTP	$X < 10$ PE	865	6	2
	$10 \leq X < 50$ PE	0	10	0
	$50 \leq X < 500$ PE	0	10	0
	$500 \leq X < 2000$ PE	0	20	5
	$X \geq 2000$ PE	0	0	1
	Total	865	36	8
Length of pressurised sewer (m)		0	480	5590
Length of gravity flow sewer (m)		0	11,700	11,620
Number of pumping stations		0	2	22
Number of manholes		0	250	350

gravity sewer. Additionally, 350 manholes and 21 pumping stations would have to be built.

Table 4.5 shows clearly the trade-off between the different scenarios where onsite wastewater management avoids the need to use sewer but increases the number of WWTPs to be built. On a technical level, this trade-off is easy to assess as the higher the number of WWTPs and pumping stations to be built, the higher the risk of failure even if the failure of one single onsite system should have a very limited impact. However, on an economical or organisational basis, operating and maintaining a high number of WWTPs and pumping stations can quickly become challenging and expensive (energy costs, spare parts, pumps' replacement). For the small-scale treatment technologies robustness and low O&M requirements are usually key selection criteria when deciding which technology to use. These criteria typically limit the operating costs but may increase the capital to be invested at the beginning of the project. On an environmental level, onsite and small-scale robust wastewater treatment systems are usually less efficient at removing nutrients. This may be an issue in sensitive areas and may require the use of more complex and expensive (capital and O&M) treatment technologies.

Costs analysis

Table 4.6 presents the results of the costs estimations calculated according to the technical specifications presented in Table 4.5. For this analysis, the assumptions taken were similar to those taken for evaluating the urban scenarios with the exception that constructed wetlands were selected as the wastewater treatment technology in replacement of the SBR technology. The reasons are that in this rural context, obtaining land for the construction of the WWTPs is not assumed to be a constraint and that selecting a more robust but space-demanding technology is more appropriate. Constructed wetlands have the capacity, if designed and operated correctly, to reduce total nitrogen by up to 60% by performing both nitrification and denitrification. Constructed wetlands are able to reach higher treatment performance but in this context, it has been decided to be conservative. Finally, constructed wetlands have also been selected as they can be designed to be operated without the need of pumps, limiting operating costs and the risk of failure of the moving parts.

Table 4.6 (onsite wastewater management) is the cheapest option over the project life span. The main reason is that scenario 1 has lower capital costs as it does not require sewer and pumping stations. However, Scenario 1 has higher O&M costs. This could pose a problem as each owner of onsite treatment systems has to pay every year for the O&M costs while for collective systems like in scenarios 2 and 3, the O&M costs are ventilated through the water bill (a few cents per cubic meter of water consumed). While scenario 1 is the cheapest option, scenario 3 (high clusterisation) is the most expensive option with a very high level of investment required for the pressurised sewer and the pumping stations. These high investment costs coupled to higher O&M costs do not overcome the low level of investment required for the WWTPs themselves. Finally, the cost of scenario 1 (high decentralisation) is 20% cheaper than scenario 2 and almost half the price of scenario 3.

Table 4.6 Cost comparison of the rural scenarios

Costs	Station	Rural scenario 1	Rural scenario 2	Rural scenario 3
Capital costs ($)	WWTP[a]	3,304,662[b]	880,890	388,361
	Pumping stations	0	31,084	326,383
	Pressurised sewer	0	297,728	3,358,788
	Gravity sewer	0	3,190,374	3,178,610
O&M costs ($ year^{-1})	WWTP	60,550	14,870	12,325
	Pumping stations	0	130	1,661
	Pressurised sewer	0	1,668	18,813
	Gravity sewer	0	37,447	37,310
Capital costs total ($)		3,304,662	4,400,076	7,252,141
O&M costs total ($)		1,310,625	1,171,316	1,517,540
Net present value ($)		4,615,286	5,571,392	8,769,682

[a]Includes the reinvestment every 10 years for the treated effluent disposal system
[b]Does not include the creation or upgrade of a sludge treatment facility

Recommendations

While scenario 1 is the cheapest, the cost differential between scenario 1 and scenario 2 is not so high as to prevent scenario 2 to be selected. Indeed, the limitation in the number of WWTPs between scenario 1 and 2 (from 865 to 36) would simplify O&M activities while enabling more frequent visits (less sites to visit) and limit the risk of failure. Scenario 2 would thus be the preferred option. Scenario 3 is in all aspects the least preferred option as it is both technically complex (number of pumping stations and length of pressurised sewer) and expensive. The only interest of scenario 3 is to limit the number of WWTPs to seven, simplifying O&M activities.

4.6 Decentralised Wastewater Management for Pollution Alleviation At Catchment Scale

4.6.1 Evaluating Potential Alleviation

The scenarios presented above for the two test regions have been developed according to the local specificities. Having the overall objective to intercept and treat raw wastewater before its nutrients can pollute the local environment and ultimately Chao Lake, these scenarios are mainly differentiated by their technical specifications and their economic consequences. However to improve the water quality of Chao Lake, it is important to assess the level of pollution alleviation that the different can achieve. Eventually, this assessment asks the question of how to best invest resources (money) for limiting wastewater pollution? Before comparing the scenarios together, the simple decision to modify the technical components in type or number can have a significant impact on the amount of pollutants removed. Usually, more technically complex and energy intensive systems are able to provide better nutrient removal. However, complex systems are more prone to failure and a single failure can have a more serious impact, over the long term, than a lower treatment performance. A pollution alleviation assessment of the different scenarios will be presented below. This assessment relies on the treatment capacity of the selected technologies for the common parameters of BOD_5 and TN. These two parameters are among the highest contributors to the pollution of Chao Lake and their removal mechanisms by the selected technologies are usually quite well understood. The treatment performance of the technologies used below is quite conservative to avoid over estimating the pollution alleviation potential. Other pollutants such as phosphorus have knowingly been excluded from the study as they commonly require additional treatment steps that make the systems more expensive and more likely to fail (the more treatment complexity, the higher the risks of system failure).

Pollution alleviation in urban areas

At the scale of the urban test region, the c.a. 13,000 inhabitants of the urban villages are currently responsible for the production of 384,300 m^3 of wastewater per year (water consumption of 101 litres per person per day with an estimated typical

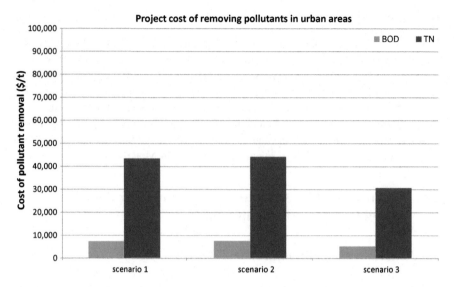

Fig. 4.13 Cost of removing pollutants in urban areas over the project life (30 years)

wastewater return ratio of 80%) containing about 285 t of BOD_5 (60 g produced per person per day) and 58 t of TN (12.2 g produced per person per day). The different scenarios created are offering a quite similar environmental performance. Indeed, it is expected that the technologies selected (SBR or connection to the activated sludge municipal wastewater treatment plant) are similarly able to remove 95% of the BOD_5 and 80% of the TN. This means that only 15 t of BOD_5 and 12 t of TN would be released in the environment each year. While the release in the environment of this amount of pollutants would still be excessive, it would offer a net improvement compared to the current situation. If the connection of the urban villages to sewer and wastewater treatment was to happen, it is to be hoped that the natural remediation capacity of the Shuangqiao River combined to some river water quality improvement infrastructure currently built would contribute to greatly limit the flow of pollutants towards Chao Lake. In terms of costs over the life span of the project (30 years), Fig. 4.13 highlight the fact the treatment of BOD_5 is always cheaper than the treatment of TN. Additionally, the treatment costs of scenario 3 are also cheaper than those of scenarios 1 and 2. This is due to the fact that in scenario 3, the treatment of the wastewater from the 6600 PE connected to the municipal WWTP is assumed to be cost neutral (no upgrade required).

Pollution alleviation in rural areas

At the level of the rural test region, it is expected that the estimated 2600 inhabitants are currently responsible for the release into the environment of 77,000 m^3 of wastewater per year containing c.a. 57 t of BOD_5 and 11.6 t of TN. The final selection of a rural scenario has a strong impact on the potential pollution alleviation due to the specific technological assumptions taken. In scenario 1, it is assumed that

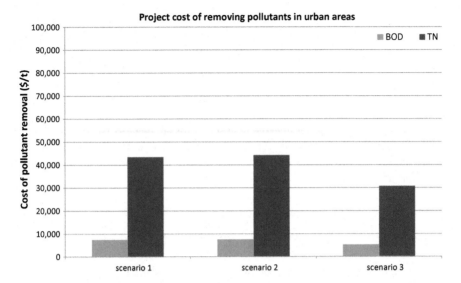

Fig. 4.14 Cost of removing pollutants in rural areas over the project life (30 years)

one septic tank and one dispersal fields would be used for each building in the test region. Sceptic tanks can be expected to reduce the BOD_5 discharge by c.a. 65% with some of the remaining BOD_5 as well as a little bit of the nitrogen removed through soil passage and plant uptake in the dispersal field. However, it is still expected that relying on onsite treatment could release up to 11.5 t of BOD_5 per year and 10 t of TN. In the case of scenario 2 and 3, the use of constructed wetlands enables a much better protection of the environment. Indeed, constructed wetlands can be expected to remove at least 90% of the BOD_5 load and between 40 and 60% of the TN load, according to the design selected. In the case of your test region, this result on less than 6 t of BOD_5 and between 6.6 and 4.2 t of TN released in to the environments. Another interest of scenario 2 and 3 is that the collection and treatment of all the wastewater from a village in a single place enables the setting up of irrigation reuse schemes. These schemes would enable the implementation of nutrient management plans that could reduce the use of fertilisers and contribute to reducing non-point source pollution which is much more complex to halt than wastewater pollution. On an economic level over the life span of the project (30 years), it is interesting to observe on Fig. 4.14 that scenario 1 is the cheapest in terms of BOD_5 removal but the most expensive in terms of TN removal. This is due to the fact that the robust onsite treatment technology selected (septic tank and disposal field) is not the optimal solution for managing nitrogen. Scenarios 2 and 3 have a BOD_5 removal cost that are similar but scenario 3 is more expensive to remove TN. This is explained by the fact that the cost of the sewer network brought back to the amount of TN removed is higher in scenario 3 than in scenario 2. Scenario 2 would thus be the most cost efficient option in rural areas for removing pollutants.

Finally it is interesting to note that rural scenario 2 is overall cheaper than the two of the three urban scenarios. This is uncommon but can be explained by two factors. The first factor is that the rural scenario deals with only 2600 PE while the urban scenarios deal with 13,000 PE. This puts a bias in the calculations as the numbers obtained are based on the treatment performance of the selected technologies expressed in percentage removal of the total load of pollutant and not brought back to the inhabitant. Factoring in the number of inhabitants connected would have shown that urban wastewater management scenarios are cheaper per unit of nutrient removed than the rural scenarios. The second reason is that in urban areas, the lack of space requires heavier sewer infrastructure (with more pumps) that increase the costs for the scenario.

4.6.2 Replication Potential in Catchment of Chao Lake

Having understood the potential pollution alleviation of the scenarios for each test area, it is important to evaluate the potential replication of decentralised wastewater management in the catchment of Chao Lake. First of all, the catchment of Chao Lake covers 13,130 km^2. In this area are living 11 million inhabitants with at least. 3.2 million in urban area and 7.8 in rural areas (Asian Development Bank, ADB 2015). However, this population repartition is likely to change as several projects are under way to relocate some rural population towards urban centres (Asian Development Bank, ADB 2015) and some urban population to make place for new infrastructure and improve quality of life (Asian Development Bank, ADB 2017). These resettlement plans have a serious impact on the potential implementation of decentralised wastewater management in Chao Lake catchment as not enough data is available to clearly assess which criteria will be used to decide of the potential transfer of population. The sub-chapters below present in more details the context and potential for replication of decentralised wastewater management in urban and rural areas.

Replication potential in urban areas

Concerning the urban area, the implementation of decentralised wastewater management may be feasible in several situations. The first situation would be the urban areas where the local authorities have decided to not replace the existing unconnected buildings with newer developments. In these areas, decentralised wastewater management could be an option albeit probably more expensive than connecting these areas to existing municipal WWTPs. This situation would require the local authorities to accept the responsibility and risks associated to have small WWTPs located within the urban area (municipal WWTPs are usually located outside urban area). A second situation would be to apply decentralised wastewater management to new urban area under development in order to avoid overloading or upgrading existing centralised WWTPs. This option would also have the benefit of enabling new economic activities relying on water availability (urban farming, cooling for industry, etc.). An important aspect of this second point would also be to spread the

Fig. 4.15 Photos of failing wastewater treatment system in a rural village (**a**) and (**b**)

release of treated wastewater to different locations to improve the local water cycle. Centralised municipal WWTPs are usually collecting (waste)water over a wide area while releasing it in a single location, unbalancing the local water cycle. The potential for replicating decentralised wastewater management in the urban area of Chao Lake Catchment is thus expected to be reduced. It could however be an interesting option for the transition phase of fast urbanisation that China is experiencing.

Replication potential in rural areas

If some rural villages are quite clearly left to "die", it is unlikely that all rural villages will be abandoned as this would hinder local economic activities and especially agriculture. It is thus more likely that the local authorities will encourage the aggregated of the rural population in some bigger villages where basic services (electricity, water, sanitation, etc.) will be improved. In these villages, implementing decentralised wastewater management could make sense and would have to be assessed. From our observations of the rural areas and our discussions with the local populations, few villages have been renovated so far and are equipped with wastewater collection and treatment technologies. Still, in the renovated villages, the failure of wastewater treatment systems (Fig. 4.15) has been observed and linked to inadequate technological choices (equipment unfit for the harsh wastewater environment, design errors, etc.) as well as poor or inexistent operation and maintenance activities. It is thus expected that implementing decentralised wastewater management in rural area would require the selection of more robust technologies and the implementation and enforcement of a monitoring plan with contingencies procedures established for failing systems.

As mentioned before, 7.8 million inhabitants are currently living in the rural areas of the catchment of Chao Lake (Asian Development Bank, ADB 2015). This rural population produces c.a. 632,000 m^3 of wastewater every year and c.a. 170,850 t of BOD_5 and 34,750, of TN per year. By simple extrapolation, the costs for providing decentralised wastewater management for 7.8 million people could reach between 14 billion (scenario 1) and 26 billion dollars (scenario 3). To avoid reaching such a high cost, the interventions should be prioritised and the technology standardised (one type of technology/design for a similar situation). Additionally, this estimation does not consider economies of scale that could be made by awarding design and construction contracts for tens of systems instead on individual systems. Finally, the relocation of some rural population towards new urban developments would also reduce the costs as it is always cheaper to build new infrastructure than to retrofit existing situation (opening and closing up roads for laying pipes is expensive) and as relying on centralised municipal wastewater treatment systems in these new urban developments would be more cost-efficient that building several small scale WWTP in existing villages.

4.7 Conclusions

Low clusterisation happen to be the most cost-efficient wastewater management approach in the rural area around Chaohu while a combination of clusterisation and centralisation presents the best perspectives within the urban area of Chaohu City itself. In urban areas centralised wastewater management offers significant cost savings compared to decentralised options. It may however not be feasible if the existing infrastructure is not able to cope with the additional inflow of wastewater. In rural areas, centralisation becomes an option too expensive while onsite wastewater management is unlikely to provide sufficient environmental protection to improve water quality in Chao Lake. For both urban and rural areas, the bulk of the costs are related to the sewer infrastructure while the technological decisions impact the long term O&M costs and the economic burden that is carried over the lifespan of the treatment systems. It is thus important to select the collection and treatment technologies according to the specific local needs but also to limit the technological variability to profit from economy of scale, both for the capital invested and for simplifying O&M activities. The ALLOWS tool is in a position to provide stakeholders with a solution to identify and prioritise the local/regional needs for wastewater management while at the same time estimating the financial requirements that the different options would entail. The ALLOWS tool enables the local stakeholders to build their wastewater management plan and to start the discussion with the funding agencies in order to reach sanitation and environmental protection goals. Increasing the level of technological specification (phosphorus removal, emerging pollutants, etc.) during the scenario creation and comparison would also provide the stakeholders with additional data that could be used to prioritise the investment of available resources towards more urgent pollution sources or polluted areas. Finally, an iterative process using the ALLOWS tool to evaluate the local policies such as population relocation and adapting these policies according to the results provided by the tool would contribute to take more cost-efficient decisions but also decisions that would provide a better environmental impact.

References

Asian Development Bank, ADB. *Reviving Lakes and Wetlands in the People's Republic of China, Volume 2: Lessons Learned on Integrated Water Pollution Control from Chao Lake Basin.* Asian Development Bank, ADB (2015)

Asian Development Bank, ADB. The Water Environment Improvement Engineering of the City Proper of Chaohu City of Water Environment Improvement Project of Anhui Chao Lake Basin Loaned by Asian Development Bank (Mid-term Adjustment) (2017). https://www.adb.org/sites/default/files/project-documents/44036/44036-013-rp-31.pdf

van Afferden M, Cardona JA, Lee M-Y, Subah A, and Müller RA. A new approach to implementing decentralized wastewater treatment concepts. Water Sci. Technol. **72**(11), 1923–1930 (2015)

Chapter 5
WP-C: A Step Towards Secured Drinking Water: Development of an Early Warning System for Lakes

Marcus Rybicki, Christian Moldaenke, Karsten Rinke, Hanno Dahlhaus, Knut Klingbeil, Peter L. Holtermann, Weiping Hu, Hong-Zhu Wang, Haijun Wang, Miao Liu, Jinge Zhu, Zeng Ye, Zhaoliang Peng, Bertram Boehrer, Dirk Jungmann, Thomas U. Berendonk, Olaf Kolditz and Marieke A. Frassl

5.1 Introduction

Marcus Rybicki, Marieke Frassl, Dirk Jungmann, Karsten Rinke

Lakes are important ecosystems that provide a number of ecosystem services including provision of drinking water, flood control, fisheries and in general a high natural, cultural and aesthetic value. Provisioning services from lakes are particularly relevant in regions where lakes supply drinking water. In these water bodies, a high water quality is of utmost importance in order to produce drinking water at required quantities and at affordable prices. High nutrient loading, eutrophication, and toxicant pollution, however, are growing stressors in many places, driving severe water quality deteriorations that harm domestic water supply, quality of life and social welfare.

M. Rybicki (✉) · D. Jungmann · T. U. Berendonk
Department of Hydrosciences, Institute of Hydrobiology, Chair of Limnology,
Technische Universität Dresden, Zellescher Weg 40, 01217 Dresden, Germany
e-mail: marcus.rybicki@tu-dresden.de

D. Jungmann
e-mail: dirk.jungmann@tu-dresden.de

T. U. Berendonk
e-mail: thomas.berendonk@tu-dresden.de

K. Rinke · B. Boehrer · M. A. Frassl
Department of Lake Research, Helmholtz Centre of Environmental Research–UFZ,
Brückstraße 3A, 39114 Magdeburg, Germany
e-mail: karsten.rinke@ufz.de

B. Boehrer
e-mail: bertram.boehrer@ufz.de

M. A. Frassl
e-mail: marieke.frassl@ufz.de

© Springer Nature Switzerland AG 2019
A. Sachse et al. (eds.), *Chinese Water Systems*, Terrestrial Environmental Sciences,
https://doi.org/10.1007/978-3-319-97568-9_5

Fast growing urban areas are particularly vulnerable to these deteriorations in surface water resources, because waste, waste water, and chemical pollutants (heavy metals, pesticides, etc.) are affecting nearby aquatic ecosystems. While in river ecosystems these pollution pressures only affect water users further downstream, i.e. not directly the pollution producer responsible for the water quality deterioration, standing water bodies like lakes or reservoirs directly and often negatively feed back to the adjacent urban communities.

C. Moldaenke · H. Dahlhaus
bbe Moldaenke GmbH, Preetzer Chausee 177, 24222 Schwentinental, Germany
e-mail: HDahlhaus@bbe-moldaenke.de

C. Moldaenke
e-mail: cmoldaenke@bbe-moldaenke.de

K. Klingbeil · P. L. Holtermann
Department of Physical Oceanography and Instrumentation,
Leibniz Institute for Baltic Sea Research, Seestraße 15, 18119 Rostock, Germany
e-mail: knut.klingbeil@io-warnemuende.de

P. L. Holtermann
e-mail: peter.holtermann@io-warnemuende.de

K. Klingbeil
Department of Mathematics, University of Hamburg,
Bundesstraße 55, 20146 Hamburg, Germany

W. Hu · J. Zhu · Z. Ye · Z. Peng
NIGLAS, Nanjing Institute of Geography & Limnology,
Chinese Academy of Sciences, 73 East Beijing Road, Nanjing 210008, China
e-mail: wphu@niglas.ac.cn

J. Zhu
e-mail: jgzhu@niglas.ac.cn

Z. Ye
e-mail: marcochen94@hotmail.com

Z. Peng
e-mail: zlpeng@niglas.ac.cn

H.-Z. Wang · H. Wang · M. Liu
State Key Laboratory of Freshwater Ecology and Biotechnology,
Institute of Hydrobiology, Chinese Academy of Sciences,
No.7 Donghu South Road, Wuhan 430072, China
e-mail: wanghz@ihb.ac.cn

H. Wang
e-mail: wanghj@ihb.ac.cn

M. Liu
e-mail: 1648873667@qq.com

O. Kolditz
Department of Environmental Informatics, Helmholtz Centre of Environmental Research–UFZ, Permoserstr. 15, 04318 Leipzig, Germany
e-mail: olaf.kolditz@ufz.de

O. Kolditz
Applied Environmental System Analysis, Technische Universität Dresden, Dresden, Germany

Excellent case studies of fast growing urban areas and coincidental deterioration of surrounding lake ecosystems can be found in China, e.g. at the lower Yangtze River. The large shallow lakes in this region are suffering from pollution and eutrophication caused by nearby urban areas and agricultural land-use, e.g. Lake Taihu and the urban areas of Suzhou and Wuxi or Chao Lake with the urban areas of Hefei and Chaohu city. Both lakes are used for drinking water supply and suffer from frequent harmful algal blooms impeding drinking water production. Operators of the drinking water plants do not only call for a long-term restoration of the lakes but also for a thorough water quality monitoring in order to be informed about the composition of the withdrawn raw water.

This chapter introduces a showcase how such a monitoring system can be designed, deployed, and operated to inform stakeholders at the lake. It further shows how such a monitoring system can be elaborated by adding knowledge from modelling. The underlying strategy of this approach is that a thorough information about the current state of the lake system and its dynamics is helping drinking water suppliers and other stakeholders to optimize their process engineering, short-term reaction to water quality deterioration, and long-term planing. We use the large and shallow Chao Lake in the Anhui province as an example of tailored in situ monitoring and modelling tools for water managers as well as the shallow Bao'an Lake in the Hubei province as a special case for the application of dynamic biomonitoring.

The design and implementation of an integrated online monitoring system is a very complex task that includes different aspects, each of which comes along with specific demands and difficulties.

1. *Equipment*: The selection of the right probes and devices depends on the characteristics of the aquatic water body (e.g. deep vs. shallow), its pollution pathways and pollution degree as well as the feasibility for local water managers. It further depends on the scientific question being investigated. An optimum solution requires a deep scientific understanding as well as a sufficient overview about available methods, techniques and instrumentation.
2. *Data collection*: This aspect contains the connection of the instruments, the collection of data and the secure transmission to an environmental data base. Nowadays, connecting the instruments and ensuring a secure transmission of data seems to be an easy task considering the progress in information technology. However, the diversity of instrument manufacturers, a varying implementation of proprietary and open data protocols as well as the diversity of available interfaces for data transmission regularly lead to conflicts and incompatibilities, which hamper the establishment and efficiency of integrated monitoring systems. To overcome these difficulties a deep technical understanding is required.
3. *Data handling*: This aspect covers the processing, storage and administration of the acquired data. Online monitoring often generates large amounts of data, due to a high temporal resolution of measurements and a large number of available parameters. The standard data user, e.g. in an authority or scientific institute, needs to focus on the scientific interpretation of the data and should be kept free from preprocessing and administering large data sets. Some of the challenges in this

aspect are the processing of raw data, which includes quality assurance and quality control (QA/QC), e.g. handling data drift, flagging of outliers, etc. Furthermore, large data sets require the facilitation of a suitable storage and backup-system, including meta information like geographical data and maintenance protocols. Finally, an easy and fast access to the whole data set needs to be implemented, including tools for data consolidation and visualization as well as automated data processing to allow for additional functionalities like early warning systems. This variety of challenges reveals the complexity of data management, which requires specific knowledge and experience in applied computer sciences as well as a close collaboration of stakeholders, manufacturers and scientists.
4. *Data utilization*: The last aspect is the data usage, e.g. for modelling and prediction purposes. Modelling serves as a tool to understand the complex interactions within an aquatic ecosystem and can therefore inform management on optimization of procedures and through scenario simulations on proactive options for action. This aspect is closely related to the type of data acquired (the equipment of the monitoring system).

Though not dealt with here, we want to stress that data security is of major importance.

To create an efficient monitoring system that considers the mentioned requirements, a complex project consortium was built, including scientific partners from the Technical University Dresden, Institute of Hydrobiology (TU Dresden) and the Helmholtz Center for Environmental Research (UFZ), Department of Lake Research, as well as commercial partners with expertise in monitoring techniques (bbe Moldaenke), data transfer (AMC) and data management (WISUTEC) on the German side. The structure of the consortium and linkage between different partners within this project is visualized in Fig. 5.1. Support on the modelling aspects was facilitated through a collaboration with the Leibniz Institute for Baltic Sea Research (IOW). On the Chinese side, the collaboration included scientists from the Nanjing Institute of Geography and Limnology (NIGLAS) as well as scientists from the Institute of Hydrobiology of the Chinese Academy of Sciences in Wuhan.

5.2 Towards an Environmental Information System for a Drinking Water Delivering Lake

Marcus Rybicki, Karsten Rinke, Marieke Frassl

We designed the components of the environmental information system for lake ecosystems to fulfil the following three aspects:

1. Conception of in-situ monitoring of physico-chemical and biological water quality variables in the lake with a particular emphasis on algal bloom detection. The goal was to make the monitoring data available by online data transfer to a central server so that real-time data can be supplied to users. These features were realized

5 WP-C: A Step Towards Secured Drinking Water: Development …

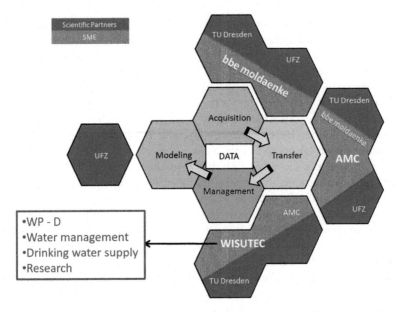

Fig. 5.1 Main tasks of the sub project WP-C, linkage between the scientific and economic partners within this work package and the respective contributions

by developing a water quality monitoring buoy with a suite of sensors, a central logger unit, and data transfer infrastructures.

2. Explicit water quality monitoring in terms of biomonitoring using a *Daphnia*-Toximeter to enable the rapid detection of acute toxic events, e.g. caused by sediment resuspension induced release of toxic compounds, algal blooms or other contaminants introduced from the catchment. The long-term objective was to test the suitability of this technique for the process control during drinking or process water treatment as well as for general surface water monitoring purposes in China.
3. Establishing a three-dimensional hydrodynamic model of Chao Lake in order to characterize temperature dynamics, transport and currents as well as bottom shear stress as a proxy for the occurrence of resuspension events.

5.3 Study Sites: Chao Lake and Lake Bao'an

Marieke Frassl, Marcus Rybicki

Chao Lake is a shallow, large lake having a surface area of about 780 km^2 and a relatively small average depth of only 3 m (Kong et al. 2017; Zan et al. 2011). The lake suffered from severe anthropogenic pressure from structural degradation, nutrient pollution and eutrophication (Kong et al. 2017). For many years now, this has resulted in the frequent occurrence of massive cyanobacterial blooms. Besides high external phosphorus inputs (Kong et al. 2017), internal loading is extraordinarily high being maintained by high phosphorus contents within the sediments (Zan et al. 2011). While Chao Lake is intensely polluted from the province capital city, Hefei, whose wastewater drains into the lake's western basin, Chaohu city on the eastern shore of the lake still uses the lake as a drinking water source. The occurring massive algal blooms are a major problem for the drinking water production there (Fig. 5.2).

Lake Bao'an is located at the south part of the middle reaches of the Yangtze River in the east of Wuhan (Hubei Province). The lake is a representative medium sized shallow floodplain lake of the Yangtze catchment (40 km^2, Wang et al. 2014a) and can be classified as a sub-urban lake. It is surrounded by many fish ponds used for aquaculture. Furthermore, the catchment of the lake comprises urban areas with the city of Bao'anzhen in the south as well as agriculturally used areas on the western side. Lake Bao'an used to be clear, with a diverse macrophyte community in the past. Due to inputs from aqua- and agriculture as well as from urban areas during the past decades water quality has deteriorated, causing a reduction of the macrophyte diversity. Today the lake's water quality does not meet the required standards (Environmental quality bulletin of Hubei Province in 2016), especially regarding total phosphorus and further pollutants like persistent organic pollutants (POPs). The Institute of Hydrobiology of the Chinese Academy of Sciences runs a mesocosm facility in the east of the lake, which was utilized for biomonitoring purposes and includes various ponds with specific physico-chemical characteristics (Wang et al. 2017).

Fig. 5.2 Scum-forming cyanobacteria and further signs of intense pollution on the shore of Chao Lake (Source: K. Rinke)

5.4 Materials and Methods: Water Quality Monitoring, Biomonitoring and Lake Modelling Tools

Marieke Frassl, Hanno Dahlhaus, Christian Moldaenke, Knut Klingenbeil, Marcus Rybicki

Recent developments in sensor technology and online services enable a multi-layered monitoring of surface water resources. This follows a multi-barrier approach meaning that water quality is observed in situ, i.e. within the lake ecosystem, and along water abstraction or drinking water production. To allow immediate management interventions and proactive action, it is optimal to monitor the water at these different locations (before and after abstraction) as well as establishing an online-system, which collects those data in one location and provides real-time information to water managers and scientists. Therefore, one of the most important features of the proposed monitoring concept was the online capability of the used probes and loggers.

Our monitoring strategy for urban lakes included three components: 1st – Monitoring of in situ water quality dynamics within the lake based on an autonomous buoy, hosting sensors, energy storage, logger technology and online data transmission (Sect. 5.4.1); 2nd – Effect-monitoring of potential toxic effects in abstracted surface waters based on biomonitoring technologies, which is a technique particularly suited for water quality monitoring within water infrastructures, e.g. drinking water plants (Sect. 5.4.2); 3rd – An integrated data platform used for storing monitoring data in a database, visualizing recent trends or status quo values, and for organizing a structured data transfer between the monitoring stations and the data platform (Sect. 5.4.3).

We complemented our monitoring efforts by building a lake model that enabled us to simulate water transport and hydrodynamic characteristics of the lake (Sect. 5.4.4). This was achieved through the application of a three-dimensional hydrodynamics model that can be used to understand dominant flow patterns, stratification dynamics and the occurrence of resuspension events.

5.4.1 In-situ Water Quality Monitoring

5.4.1.1 Water Quality Monitoring Buoy for Meteorological and Water Quality Variables

We used an established buoy system, that has been used widely within Germany, and modified it together with a small engineering enterprise (ENVIMO GmbH, www.envimo-gmbh.de) in order to develop a robust and reliable modular system for water quality monitoring. This buoy system, called AWATOS (Autarkic Water Observation System) is manufactured by ENVIMO and now commercially available. The technical features of the AWATOS buoy can be summarised as follows:

- Minimised energy consumption by logger and sensors combined with solar panels, wind generator and large battery capacities for long-term operations without external electricity supply (the system was successfully tested during German winter conditions with long overcast periods, short periods of daylight and low sun elevation).
- Central logger unit with flexible data import protocols (SDI-12, analogue, RS-485, Modbus) and an online data transmission via telecommunication networks and server infrastructures (e.g. by FTP).
- Three-point anchorage that minimises rotation of the buoy body; a compass unit tracks the rotational movements of the buoy to allow for correction of systematic errors in wind direction measurements induced by remaining buoy rotations.
- A central rod to facilitate meteorological measurements 2 m above the lake surface.
- Four vertical tubes enable the deployment of maximum four sensor probes; an additional rack allows for the deployment of one larger probe underwater; flexible data exchange protocols of the internal logger enable the integration of most sensor and probe systems available on the market.

We equipped the monitoring buoy with two water quality probes and a number of meteorological sensors. For the latter, air temperature, wind speed and direction, humidity, air pressure, and precipitation were measured by a combi-sensor (Vaisala WXT 520). Global radiation was measured by a pyranometer (SMP-11-A, Kipp & Zonen). The water quality probes were a multi-parameter probe (EXO2, YSI) for basic physical, chemical and biological variables and a multi-channel fluorescence probe (Phycoprobe, bbe moldaenke) for a more detailed observation of algal community dynamics. Both probes are described in more detail in separate sections below.

There are different techniques to deploy the buoy into the water. It has hooks that allow to lift the buoy with a crane and lower it into the water. However, this requires a position close to the shore where the water level is deep enough to let the buoy float, e.g. next to a bridge. This procedure allows to set up the whole equipment while being ashore (see Fig. 5.3).

After a first test phase in a German reservoir the buoy was shipped to China and deployed in Chao Lake in the north-eastern bay near the lake outflow. This position is relevant to the observation network as it monitors that lake segment, which is used by the water infrastructures of Chaohu city. Data were stored on the internal logger (netDL 1000, OTT Hydromet GmbH) and sent to an external FTP server twice per day. The collected data would be assembled in a database and provided to the users through a specialised software interface (ALVIS software, Wisutec GmbH, see below and Chap. 6). Though the buoy in principal is a self-conntained system (all sensors underwater automatically are wiped prior to measurement or at least an hourly interval), it still requires regular maintenance, like cleaning or calibration of the sensors. We successfully worked with a maintenance interval of four weeks.

Fig. 5.3 Equipping the buoy for deployment in Chao Lake in March 2017 (Source: Z. Ye)

5.4.1.2 Multiparameter Probe for Basic Water Quality Variables

We deployed a YSI EXO2 multi-parameter probe in one of the vertical tubes of the buoy, i.e. at approximately 0.75 m depth, directly beneath the body of the buoy. This probe is suitable for long-term deployments, because an automatic central wiper can be installed at the sensor head. This wiper cleans the sensor surfaces in regular time intervals (e.g. once per hour or before measurement), it avoids biofouling of sensors in a highly efficient way and with minimum energetic costs (Fig. 5.4). In case of Chao Lake, operation over more than 10 months documented the reliability of the wiper system. The EXO2 probe can be equipped with various sensors. Due to a modularized system, the user can connect the sensors of interest and change sensors if required. During our deployment, the probe was equipped with sensors for temperature, electrical conductivity, pH, oxygen, turbidity, total chlorophyll *a* fluorescence, and phycocyanin fluorescence.

5.4.1.3 Multi-channel Fluorescence Probe for Monitoring Algal Community Dynamics

We used a newly developed multi-channel fluorescence probe called PhycoProbe – a new member of the bbe FluoroProbe line of chlorophyll fluorometers built by the bbe Moldanke GmbH (see Fig. 5.5). These instruments excite photosynthetic pigments

Fig. 5.4 Deployment of the AWATOS buoy within Chao Lake (**a**) and view on the sensor surfaces and the central wiper of the EXO2 probe (**b**). The wiper is obviously highly efficient in avoiding any biofouling on the sensor surfaces. This picture is taken after 3 months of exposure in the lake (Source: K. Rinke)

Fig. 5.5 Side view of the new PhycoProbe (Source: bbe Moldaenke GmbH)

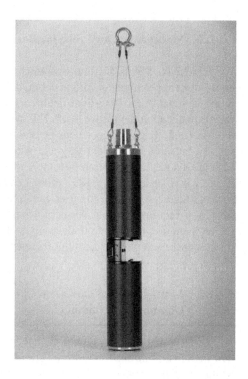

in cyanobacteria and higher algae and with specific wavelengths in separate. The resulting fluorescence signal for each exciting wave length of the central chlorophyll a is then detected and attributed to the corresponding pigments. Standard conversions were used to calculate pigment concentrations from fluorescence.

Depending on their light accessory pigments, microalgae can be divided into different 'spectral groups'. The bbe FluoroProbe differentiates between those different spectral algae groups by using light emitting diodes as light sources at five distinct wavelength: 470, 525, 570, 590 and 610 nm. The measured fluorescence emission at a wavelength band around 680 nm is then translated to four spectral algal groups: green algae (detection pigment: chlorophyll b), cyanobacteria (detection pigment: phycocyanin), brown algae (diatoms and dinoflagellates, detection pigment: fucoxanthin), and cryptophytes (detection pigment: phycoerythrin).

The chlorophyll content per algal group is estimated mathematically through fitting the normalized excitation spectra per algal group into the measured fluorescence signal. The result is the total chlorophyll concentration of the sample apportioned into the individual concentrations per spectral algal group. To avoid falsified readings by yellow substances, e.g. degradation products of biomass, the instrument contains one UV-LED in the UV-A band (370 nm) for correcting for fluorescence signals from yellow substances.

Unlike the bbe FluoroProbe, the former version of this probe, the bbe PhycoProbe is equipped with a second sensor for fluorescence. It detects the emission spectrum of extracellular phycocyanin at 650 nm, which will be excited at 590 nm/610 nm. Differentiating this emission signal from the other fluorescent signals delivers additional information about the physiological condition of cyanobacteria complementary to the concentration information via the chlorophyll content. Laboratory tests showed a close relationship between the release of phycocyanin from the photosystem within the cell prior to the toxin release during subsequent cell lysis. The signal of unbound phycocyanin can therefore be used as a proxy for extracellular toxin concentration making it highly valuable for the management of harmful cyanobacterial blooms. This is particularly relevant for the drinking water production since extracellular toxins are far more difficult to remove from the raw water than cell-bound toxins.

In the scope of this project the new bbe PhycoProbe was installed as part of the monitoring buoy and was therefore adapted to long-term deployment. Power supply was provided directly by the buoy instead of utilising an own battery. This resulted in a significantly prolonged operation period compared to standard deployment. The instrument was furthermore equipped with an automated cleaning system, the instrument-specific HydroWiper system (Zebra Tech Ltd. NZ), which is available for both fluorimetric systems of bbe, i.e. the PhycoProbe and the FluoroProbe. For the buoy application a pivoted mounting frame was developed. That way the PhycoProbe can be easier accessed from the side of the buoy during maintenance. When submersed, the instrument rests underneath the buoy's body, so that the shading effect by the buoy reduces potentially interfering direct sunlight on the sensors. For maintenance the frame can be folded up and the PhycoProbe can then be fixed at the side of the buoy, outside the water (see Fig. 5.3).

5.4.2 Biomonitoring and Associated Monitoring Technologies

5.4.2.1 Integrated Detection of Acute Toxicity Using a Dynamic *Daphnia*-Toximeter

The usage of animals as monitors for hazardous environmental conditions has a long tradition, e.g. with the utilisation of canary birds in subsurface mining. This principle of integrated organism-based monitoring is nowadays used in online toximeters like the *Daphnia*-Toximeter. In contrast to chemical analyses, where selectively substances are qualified and quantified, biomonitors expose test organisms to water, e.g. surface or process water, and a software algorithm then uses proxies, like the survival and behaviour of these test organisms, to assess the current pollution of the water. Hence, biomonitors do not detect a specific substance, but evaluate the integrated toxicity of the tested water in close relation to the time of exposure (online). This enables the rapid detection of toxic chemical spills with unknown composition and the initiation of subsequent measures, e.g. to identify the polluter. Since the beginnings of this technology in the 1970s the technique advanced markedly, increasing its sensitivity and applicability. Nowadays, dynamic biomonitors are offered with different organisms and sensitivities and are regularly used in Europe for surface water monitoring.

Within this project the sensitive *Daphnia*-Toximeter (version 2) from bbe Moldaenke was chosen to monitor surface waters used for drinking water supply (Fig. 5.6). *Daphnia*-Toximeters are known to react quickly and sensitive to toxic compounds (Green et al. 2003). Even odour compounds like β-cyclocitral and 2(E),4(E),7(Z)-decatrienal can be detected using *Daphnia*-Toximeters (Watson et al. 2007), making them an innovative tool for online water monitoring.

Daphnia-Toximeters are originally based on the acute *Daphnia magna* toxicity test according to ISO 6341. In contrast to the static ISO test, the toxicity of the water is continuously determined by observation of the swimming behaviour of daphnids, which is recorded by CCD-cameras. An online image analysis software estimates the swimming paths of the daphnids and calculates different parameters, which characterise their current behaviour (e.g. average swimming speed or average swimming distance). Specialized software algorithms (detectors) are monitoring the progression of each of the different behavioural parameters and generate penalty values of specific weight and duration in case of significant alterations. All penalty values are summed up to the so-called "toxic index" (TI), which constitutes an integrative parameter of the current behaviour and, hence, a measure of water toxicity. If the TI reaches pre-defined thresholds an alarm is triggered and further measures can be initiated. A detailed description of the *Daphnia*-Toximeter can be found in Lechelt et al. (2000).

5 WP-C: A Step Towards Secured Drinking Water: Development …

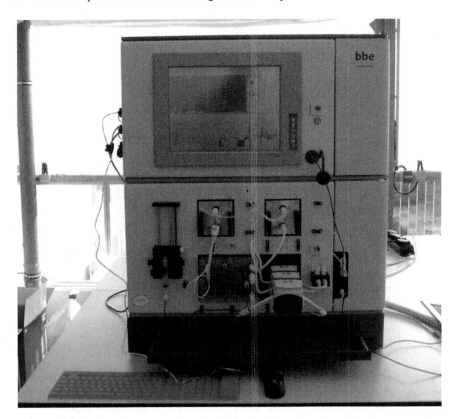

Fig. 5.6 The used *Daphnia*-Toximeter (bbe Moldaenke) during a test at the Institute of Hydrobiology, (TU Dresden, Germany) (Source: M. Rybicki)

5.4.2.2 Multi-channel Fluorescence Probe for Monitoring Fractions of Dissolved Organic Carbon

Besides phytoplankton the water matrix contains biopolymers like proteins and other diluted organic substances, which can be detected via fluorescence measurements. Humic and fulvic substances are one of the major components within this group. They are natural substances that result from the breakdown of plant and animal biomass. Such diluted substances are an important parameter to be assessed in water monitoring. Although they are not toxic, they play an important role in the treatment of drinking water. The utilisation of ozone in the treatment process leads to the formation of biologically easily available break-down compounds that could promote the subsequent development of bacteria in the water. The utilisation of chlorine for disinfectant purposes of drinking water on the other hand leads to the build up of chlorine-organic compounds, disinfectant by-products, like trihalmethanes (THMs).

Fig. 5.7 FluoSens 3D fluorometer from bbe Moldaenke

Several recent studies (e.g. Wagner et al. 2016) observed an increase in humic substances concentration in surface waters of the northern hemisphere, showing that future monitoring of these diluted organic compounds is increasingly important for ensuring a safe drinking water supply.

The FluoSens (bbe Moldanke, see Fig. 5.7) is a new UV/VIS 3D fluorometer, which was developed for the detection of diluted organic compounds in the water matrix. Based on the technology of other bbe fluorometers, it integrates LEDs with distinct wavelengths for excitation light. The fluorescence is detected by silicon photomultipliers on a set of distinct wavelengths. The excitation emission matrix for fluorescent diluted organic matter in typical surface water shows the grid of excitation and emission wavelength used in the bbe FluoSens (Fig. 5.8). Excitation wavelength are 245, 255, 280, 315, 430, 515 and 610 nm. The emission is detected at 328, 429, 511 and 700 nm. A typical application for this instrument is the monitoring of fluorescent substance concentrations in surface water bodies.

The FluoSens is equipped for the detection of humic/fulvic substances, proteins and algal chlorophyll *a*. It distinguish between 'high-molecular humic acids', 'low molecular humic acids', proteins (normalised to bovine serum albumin) and algal pigments of the groups of chlorophytes, cyanobacteria and diatoms.

Within the project the bbe FluoSens was deployed in a wastewater treatment plant in Kreischa (Germany), where preliminary tests for the data acquisition and logging system together with the bbe *Daphnia*-Toximeter were performed over several months.

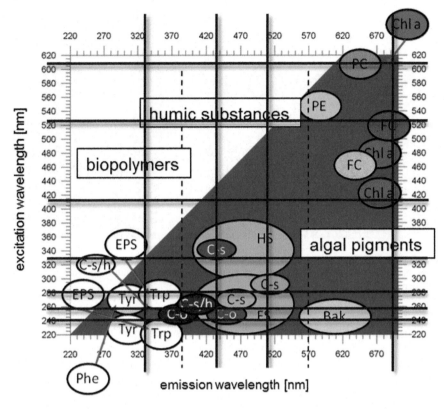

Fig. 5.8 Excitation-Emission-Matrix (EEM) for water with typical fluorescing substances. The bbe FluoSens detects humic substances, proteins and algal pigments (chlorophyll *a*) specifically. The grid shows the implemented excitation and emission wavelength according to Wagner et al. (2016)

5.4.2.3 Additional Monitoring of Physico-Chemical Base Parameters

The continuous measurement of the bbe *Daphnia*-Toximeter creates temporal highly resolved data about the toxicity of the investigated water. However, a suitable interpretation of these data requires similar high resolved data of basic physico-chemical water parameters like oxygen, pH and electric conductivity, which may affect the behaviour of the test organisms. Due to the difficult environmental conditions in Chinese surface waters, a robust but flexible system was required to adjust the physico-chemical measurements to the requirements of the biomonitoring. Additionally, the operation and maintenance by non-scientific personnel on site demanded an easy application and an easily understandable user interface to enable the correct operation of the system in China. After comparison of different suitable systems, the IQ SensorNet System of WTW (Germany) was chosen, which is a fully digital modular measurement network system designed for monitoring during waste water treatment

Fig. 5.9 The IQ SensorNet System (WTW) consisting of terminal and controller (left) as well as probes for conductivity, pH and oxygen (right) installed at the Institute of Hydrobiology (TU Dresden, Germany) (Source: M. Rybicki)

and for industrial applications (Fig. 5.9). The system for up to 20 main parameters consisted of a controller (MIQ/MC3-MOD) and a movable terminal (MQ/TC 2020XT) and was equipped with probes for electric conductivity (TetraCon 700IQ), pH (SensoLyt 700IQ) and oxygen (SC FDO 700). Additionally, an ion selective probe (VARION PLUS 700IQ) was provided by the Chinese partner to determine nitrate and ammonium concentrations.

5.4.3 Data Collection, Transfer and Management

Data processing within this sub-project was organized according to the project structure described above (Fig. 5.1). Selection of probes and instruments based on the scientific questions, and testing of innovative prototypes was done in collaboration between the scientific partners from TU Dresden, UFZ Magdeburg and bbe moldaenke. Probes and instruments were linked with two different systems: on the level of the buoy with a buoy specific data logger from OTT Hydromet GmbH (NetDL 1000) and on the level of the monitoring station using the SensoMaster32 software of AMC. Data transfer was realised via mobile services using the ftp protocol and file upload to the project server (buoy) or directly using the SensoMaster32 interface to the project database AL.VIS/Timeseries, provided by WISUTEC. Detailed information on the data processing can be found in Chap. 6.

5.4.4 Model System for Three-Dimensional Hydrodynamics and Associated Distributed Whole-Lake Monitoring

In order to complement the sparse information available from measurements, numerical simulations were carried out. Our focus on lateral gradients and circulation patterns rendered one-dimensional lake modelling (e.g. Klingbeil et al. 2018b) insufficient. Therefore, a three-dimensional numerical model of Chao Lake was set up. Three-dimensional models predict not only temperature and stratification dynamics but also lake-wide circulation patterns and transport. Such information can be valuable for lake managers, e.g. for understanding large-scale wind-driven transport of scum-forming cyanobacteria, pollutant distribution or the occurrence of resuspension events. In this project, numerical simulations were performed with the hydrodynamic model GETM (General Estuarine Transport Model; Burchard and Bolding 2002a).

5.4.4.1 The Three-Dimensional Hydrodynamic Model *GETM*

GETM solves the three-dimensional Reynolds-Averaged Navier-Stokes Equations under the Boussinesq approximation on a structured finite-volume grid. Its numerical techniques have been developed for the specific demands of coastal ocean applications (see e.g. Klingbeil et al. 2018a), but as discussed below, these are also beneficial for simulations of lakes and reservoirs. GETM facilitates the efficient calculation of a free surface with drying-and-flooding capability (Burchard et al. 2004) and applies stable integration schemes for transport in shallow areas, also required for many reservoirs with strong water level variations. Vertically adaptive boundary-following coordinates offer high spatial resolution of the surface and bottom boundary layers as well as of variable pycnoclines (Hofmeister et al. 2010). High-order internal pressure gradient schemes significantly reduce errors usually associated with boundary following coordinates. Efficient second- and third-order TVD schemes guarantee an oscillation-free and accurate transport of sharp fronts with reduced spurious mixing (Klingbeil et al. 2014). An optional non-hydrostatic extension supports the correct simulation of internal waves also with shorter wave lengths (Klingbeil and Burchard 2013). Further features of GETM, which are important for lakes, are the physical closures of unresolved effects like turbulence and wind waves, as well as the possibility to couple GETM to biogeochemical models (e.g. via FABM, the Framework for Aquatic Biogeochemical Models; Bruggeman and Bolding 2014). An interface to GOTM (General Ocean Turbulence Model; Burchard et al. 1999) provides state-of-the-art turbulence closure for the vertical diffusivity in terms of stratification and shear (Umlauf and Burchard 2005). In addition, GETM offers the inclusion of wave effects (Stokes transport, wave-induced momentum fluxes, wave-enhanced bottom friction and turbulence), depending on mean wave properties either provided by coupling to an external wave model (Moghimi et al. 2013) or parameterized in terms of local wind data. Examples of successful applications of GETM to lakes are the

5.4.5 Three-Dimensional Monitoring: Getting a Data Set for Testing the Validity of the Model

While the monitoring buoy provides a rich source of data on the state of the lake and its water quality, the buoy at the same time gathers information from only one single observation point. The application of a three-dimensional lake model, however, demands three-dimensional observation data in order to test the validity of the model. Given the large surface area of Chao Lake of more than 750 km^2 we decided for a relatively simple and cheap monitoring protocol to observe the temperature dynamics in the lake. In order to get a better understanding of the hydrodynamics within the lake, thermistor chains were deployed at up to ten locations in October 29, 2015 (see Fig. 5.10). The locations were chosen along a transect from west to east to cover as much of the lake's area as possible. Because of the considerable shipping traffic, positions were needed where it was possible to attach the thermistor strings to anchored devices like sea marks and marker buoys.

Each thermistor string consisted of three temperature loggers (TidbiT v2 Water Temperature Data Logger, Onset Computer Corp.) that were deployed at 0.2 and 1.5 m above the sediment and 0.2 m below the surface. Due to the high loss of loggers after one month of measurements, the initial 10 locations were cut down to six locations, which were maintained over several months. Data were read out regularly and lost loggers were replaced as soon as possible. At two locations, an additional oxygen logger (HOBO Dissolved Oxygen Data Logger, Onset Computer Corp.) was deployed above the sediment in order to detect oxygen dynamics.

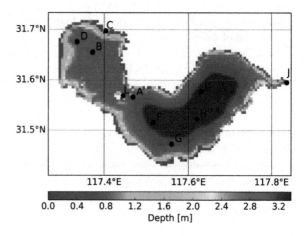

Fig. 5.10 Bathymetry of and logger positions in Chao Lake. The AWATOS buoy is deployed at position "J", close to the drinking water intake of Chaohu city

5.5 Quality Assurance for Online Measurements

Marcus Rybicki, Marieke Frassl

The thorough management of online data is an important aspect in environmental monitoring (Porter et al. 2012). Both automated routines for data storage, output and visualisations as well as data safety must be guaranteed. Professional approaches differentiate between quality assurance and quality control (QA/QC, for details see Campbell et al. 2013) and require considerable efforts. It is beyond the scope of this chapter to provide in-depth information on the applied QA/QC measures and we only briefly summarise the most important aspects according to our experience.

- Regular maintenance of sensors and monitoring infrastructures by trained staff following standardized procedures: In case of the monitoring buoy, monthly maintenance intervals are recommended. Maintenance of the buoy includes, for example, cleaning of sensors and probes, calibration of certain sensors (e.g. pH), checking the logger status and a general check of the buoy anchorage and material. In case of the biomonitoring the maintenance intervals are shorter, at approximately one week. Many manufacturers even suggest a daily inspection routine to guarantee valid measurements. Apart from the cleaning of the instruments, the exchange of the test organisms with new, fresh ones is of major importance as the entire measurements depends on the stocked organisms. Hence, for biomonitoring the cultivation or breeding of the biological material must be implemented into the QA/QC, too, which further increases the maintenance effort.
- Development of standard operating procedures (SOPs) and preparation of technical instruction sheets for the technical staff involved in maintenance or deployment and installation activities: Although manufacturers usually provide a detailed user manual, an additional SOP is useful to focus the information from the user manual to the required level of knowledge on the one hand and to provide additional information and hints useful for regular maintenance on the other hand. An excerpt from one of our SOPs on buoy deployment is provided as an example below ("How to deploy the buoy", see Fig. 5.11).
- Documentation of all performed activities on probes and instruments with information about the performed action, the performing employee and the respective timestamp: Preferably online documentation of maintenance activities into a quality-management data base is used, as it enables the real time evaluation of the status of probes and instruments in case of suspicious data. However, at least offline documentation into logbooks should be performed to enable a retrospective evaluation of the instrument status. This point is of large importance for the long-time usability of environmental data.
- Online transfer of diagnostic variables to the information system: Diagnostic variables inform about the current state of the monitoring devices and include, e.g. the voltage of the internal buffer batteries, humidity within the logger housing, or electricity supply from the solar cells. These diagnostic variables sometimes explain unexpected behaviour in the observed water quality variables, due to malfunction

Putting together the balance system/underwater weight of the buoy:

1. Mount the ballast tube with nut and washer (30 mm combination wrench). CAUTION: remove both small screws on the top of the ballast tube, they inhibit the necessary contact to the bottom of the buoy.
2. Attach the weights to the star.
3. Attach the star with the weights to the ballast tube. The orientation should be that one weight plate faces towards the bbe rack/in the middle of the solar panels.

 Useful hints: If the buoy is tilted in the water and does not stand straight, the weighing plates can be relocated.

 Useful hints: It is useful to put in the screws so that the head of the screw is underneath the screw thread/the star, that is, at the bottom. In this way, no sand will enter the screw thread, if the buoy touches the sediment.

Now you can put down the buoy a little again, so that it loosely rests on the pallet.

4. To protect the EXO-probe, mount a basket under the measuring tube of the buoy. The orientation of the basket is marked with black color.
5. The thread and the nut of the screw connection should be outside

caution: Put the EXO-probe in the measuring tube only after the buoy is in the water. If the buoy touches the sediment during deployment (for example in shallow lake parts), the basket can get in contact with the ground and may harm the probe

(credit: M.Frassl)

6. Attach the short chains for the anchorage to the anchor adapters of the buoy. The big shackle goes to the anchorage adapter, the chain with the small shackle is connected with the big shackle.
7. Attach the end of the chain with a rubber cable strap, to easily detach it in the water later.

Fig. 5.11 An example of a technical instruction sheet on the buoy deployment handed out to the technical staff in charge of this task

of sensors or data transfer. If possible, software algorithms within the monitoring devices, loggers or within the environmental information system detect critical thresholds of the diagnostic variables, e.g. low battery voltage, and generate automated email alerts that are sent to the technicians responsible for the system's maintenance.

- Regular inspection of online data in order to detect sensor drift, damages or any other kind of failures.

5.6 Development and Application of the Monitoring Systems

Markus Rybicki, Weiping Hu, Haijun Wang, Dirk Jungmann, Marieke Frassl

5.6.1 High-Frequency In situ Observation of Meteorological and Water Quality Variables in Chao Lake

The buoy was deployed in Chao Lake on March 22, 2017 by a group of researchers and technicians from NIGLAS. Preceding the deployment, a one-week training on setting up and maintaining the measurement platform took place (see Sect. 5.5). After deployment, the AWATOS-buoy continuously operated over one year without major failures except for a broken oxygen sensor in summer 2017, which was replaced within 2 months and a short logger outage in October leading to a data gap of roughly one week. Aside of these problems, which appear to be normal given the exposure of a complex monitoring system in a highly polluted lake with intense ship traffic, the monitoring system worked reliably. This high reliability and the comparatively low maintenance requirements (maintenance every 1–3 months) proved the suitability of the system for outdoor exposure. Due to the cancellation of two consecutive maintenance trips in April to June 2017, the buoy remained without maintenance for about 3 months. This long duration without maintenance led to large biofouling on the probes and sensor housings and to the colonisation of one probe by shrimps (Fig. 5.12). However, sensor surfaces remained clean and in operation due to the wiper.

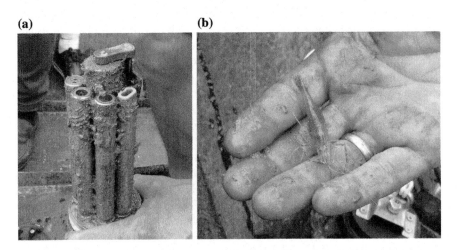

Fig. 5.12 Intense growth of biofilms and algal mats on the sensor housings due to maintenance omission (**a**) and subsequent colonisation of the housing by shrimps (**b**) (Source: K. Rinke)

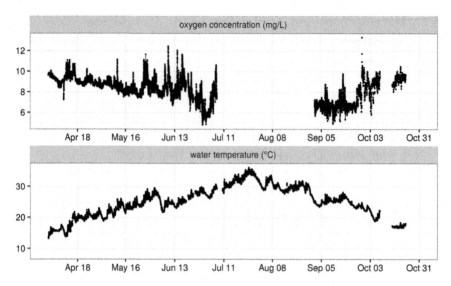

Fig. 5.13 Water temperature and oxygen concentration measured by the AWATOS buoy in Chao Lake

Over the main growing season (March until October), water temperature increased from about 12 °C to more than 30 °C reflecting a typical temperature curve of a shallow lake in the subtropical zone (Fig. 5.13). The oxygen concentration roughly followed the temperature-dependent oxygen saturating concentration, which is approximately 11 mg L^{-1} at 12 °C and 7.5 mg L^{-1} at 30 °C. Notable over- and undersaturation of oxygen have been recorded by the monitoring buoy at certain periods reflecting the eutrophic character of Chao Lake.

The continuous monitoring of the algal community dynamics by the PhycoProbe, i.e. a multi-channel fluorescence probe, is a novelty as these probes are usually employed for vertical profiling. The probe operated highly reliable and the installed wiper system for cleaning the optically relevant surfaces worked well and was stable. The algal community consisted of diatoms and green algae in spring (Fig. 5.14). In early June, i.e. at a time when surface temperatures approached 30 °C, a shift from diatoms and green algae towards cyanobacteria occurred and the community became dominated by cyanobacteria. For unknown reasons, algal biomass showed a serious drop in the second half of June inducing a clear water phase in the lake. After this period, i.e. from the beginning of July onwards, a quickly increasing biomass was recorded that was clearly dominated by cyanobacteria. During the following weeks, i.e. July and August, a massive cyanobacterial bloom occurred in the lake and the community was dominated by cyanobacteria by 70–90% (Fig. 5.14). Diatoms played no important role in the second half of the year, but chlorophytes gained importance again in autumn.

An additional and new feature of the PhycoProbe is the detection of free phycocyanin. This is a cyanobacteria-associated pigment that shows differing fluorescence properties when either bound to the photosystem within the bacterial cells or freely

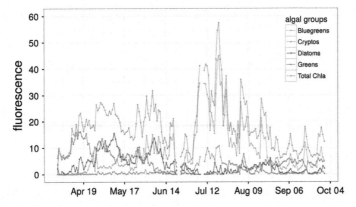

Fig. 5.14 Daily averaged fluorescence measurements by the PhycoProbe and the corresponding contributions of different algal groups to the algal community in Chao Lake

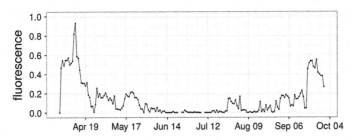

Fig. 5.15 Daily averaged fluorescence data of free phycocyanin measured by the phycoprobe in Chao Lake

dissolved in the water. A detection of this free phycocyanin, as provided by the PhycoProbe, is an indicator of cell lysis and can therefore be used as a proxy for the release of algal toxins. Our measurements in Chao Lake indicated the presence of free phycocyanin in spring and autumn (Fig. 5.15). While free phycocyanin in autumn can be attributed to collapsing cyanobacterial biomass, the high values in spring cannot be explained. More detailed field studies are required to check the contribution of resting cells from the sediment (see Ihle et al. 2005, for more details).

5.6.2 Detecting Acute Toxicity with the bbe Daphnia-Toximeter and FluoSens – Laboratory Studies with Two Model Substances in Germany

During the first phase of the project, the bbe *Daphnia*-Toximeter and the bbe FluoSens were established at the Institute of Hydrobiology of the TU Dresden in Germany. The aim was to perform measurements with model substances to test the reaction of the daphnids to chemical stress and the sensitivity of the instrument, while the data transfer from the instruments to the project data base was established and optimised.

Both instruments, especially the *Daphnia*-Toximeter, are mainly designed for usage in monitoring stations in combination with continuous water flow. Hence, successful laboratory usage under non-continuous flow conditions requires certain modifications of the instruments:

The FluoSens is equipped with a peristaltic pump (Winston-Marlow) to fill the measurement chamber with water and to exchange the sample water if required. To utilise the instrument in a non-continuous flow environment, it is sufficient to set the pumping period long enough to ensure a complete flushing of the measurement chamber with the new water sample. No further adaptation is necessary.

The *Daphnia*-Toximeter can be equipped with a special adapter for the filtration unit to enable the operation in the laboratory under non-continuous flow conditions. This modified version still requires a water volume of approximately 40 L per day. Furthermore, the supply of sufficient standardised water in laboratory experiments is difficult compared to the operation in the field, where water is directly taken from the environment. Daphnids react very sensitive to residues in tap water, especially of chlorine and copper, leading to increased mortality and invalid measurements. After testing several media, finally, modified ADaM-Medium (Klüttgen et al. 1994), which was reduced to 50% of the original contents, prepared with deionized water was chosen for the laboratory experiments. In the preliminary experiments, conducted over more then 7 days, this medium had led to the lowest mortality of daphnids.

The experiments were performed in a greenhouse of the Institute of Hydrobiology (TU Dresden, Germany). A standard water barrel (volume 150 L) was used as a reservoir for modified ADaM-media. Both instruments shared the same withdrawal tube to receive the media or the exposure solutions, allowing the instruments to run in parallel. The general schedule of the exposure experiments is shown in Fig. 5.16 and consisted of an adaption phase to stabilise the behaviour of the daphnids in the *Daphnia*-Toximeter, a short exposure phase of 6 hours and finally a post exposure phase to observe the subsequent behaviour of the daphnids over up to 4 days.

The exposure experiments were performed with the fungicide Pentacholorophenol (PCP, CAS: 87-86-5, Sigma-Aldrich, purity $\geq 98\%$), whose derivatives are still regularly used in China (Ouyang et al. 2012), and the pharmaceutical Metoprolol

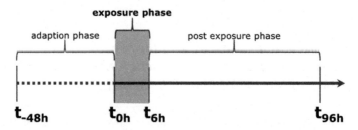

Fig. 5.16 Schedule for lab experiments with the *Daphnia*-Toximeter

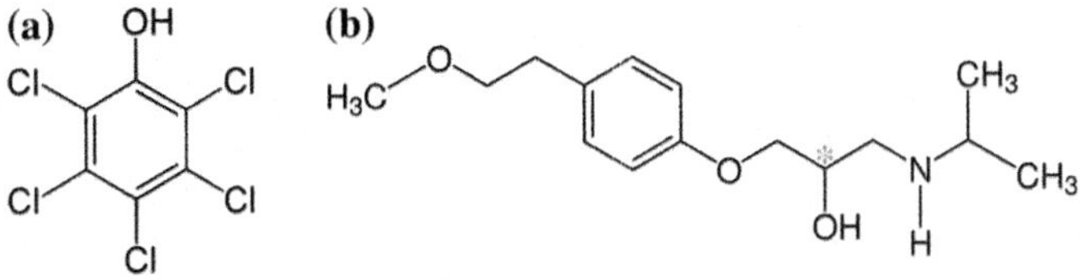

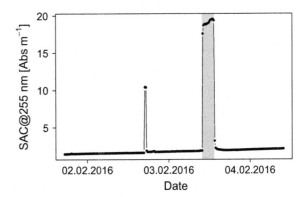

Fig. 5.17 Structural formula of the two model substances Pentachlorophenol (a) and Metoprolol (b), by Denwet [CC BY-SA 4.0] via Wikimedia Commons

Fig. 5.18 Progression of SAC255 determined with the bbe FluoSens as proxy for the PCP concentration during the experiment. The red area indicates the exposure phase with PCP. The first peak was a test with PCP stock solution

(MET; CAS: 56392-17-7, Alfa Aesar - Thermo Fischer, purity $\geq 98\%$), which is a common pharmaceutical acting as beta blocker and regularly found in European surface waters (Gurke et al. 2015, Fig. 5.17). The experiments were performed with concentrations of 5 mg L^{-1} PCP and 1 g L^{-1} MET, which is close to reported EC_{50}-values (48 h, immobility) of *Daphnia magna* for both substances. Whereas MET was solved without solvent directly in ADaM-media, PCP was solved in acetone first and then added to ADaM-media.

Exposure to Pentachlorophenol

The spectral absorption coefficient at 255 nm (SAC255) of the water determined with the bbe FluoSens was utilised to monitor the exposure during the experiment. Two distinct increases during the experiment were detected (Fig. 5.18). The first peak was a direct injection of PCP stock solution (5 mg L^{-1}) into the measuring chamber of the FluoSens to validate that the used PCP concentration was detectable with this instrument. The second peak matched well with the exposure period and therefore indicated a successful exposure of the system to PCP. The difference in SAC255 value between the validation measure and exposure is probably caused by adsorption of PCP to the tubes of the system, which may reduced the SAC255 during the validation measure. In contrast it seems plausible that the long exposure during the exposure period (6 h) caused saturation of tubes and hence a higher SAC255 signal.

PCP exposure was successfully detected using the *Daphnia*-Toximeter. The daphnids reacted quickly to the exposure by increasing firstly the swimming height

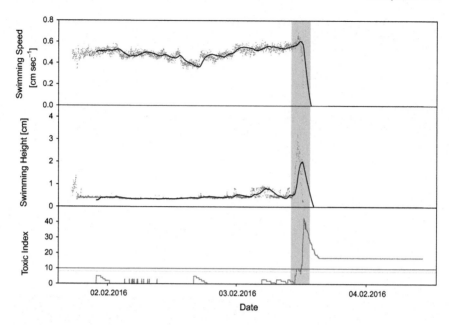

Fig. 5.19 Progression of the parameters swimming speed (upper), swimming height (middle) and toxic index (lower) during the PCP experiment in chamber 1 of the *Daphnia*-Toximeter. The red area indicates the exposure phase with PCP. Grey dots are the data points of the respective parameter with the respective prediction from the toximeter, the solid black line is based on the moving averages of 135 data points according the bbe algorithm. The calculated toxic index over all parameters is shown as a grey line in the lowest plot with the pre-alarm (yellow) and alarm (red) thresholds as horizontal lines

significantly (about 30 min after start of exposure) and subsequently also the swimming speed (about 60 min after start of exposure; Fig. 5.19). Further parameters like swimming distance, fractal dimension and the different class indices showed significant alterations, too (data not shown). In total the PCP exposure led to an increased toxic index above the pre-alarm threshold within 70 min and above the alarm threshold within 135 min. It should be noted that PCP caused total immobility of daphnids in the chambers within two hours. Hence, the experiment represents a worst case scenario with mortality of all daphnids.

Exposure to Metoprolol

As in the PCP experiment the exposure during the MET experiment was monitored with the bbe FluoSens using the SAC255 signal. The SAC255 again showed two distinct peaks of 40 abs m^{-1}, the first from a validation measure with MET stock solution (1 g L^{-1} MET) directly applied to the FluoSens and the exposure period as second peak (Fig. 5.20). Hence, the FluoSens indicated a successful exposure of the test system. Due to the physico-chemical properties of MET and the resulting low adsorptive capabilities, the SAC of both, the validation measure and the exposure period, were similar.

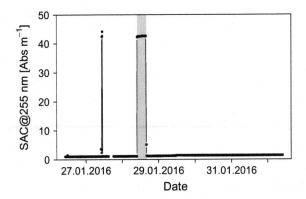

Fig. 5.20 Progression of SAC255 determined with the bbe FluoSens as proxy for the MET concentration during the experiment. The red area indicates the exposure phase with MET. The first peak was a validation measure with MET stock solution

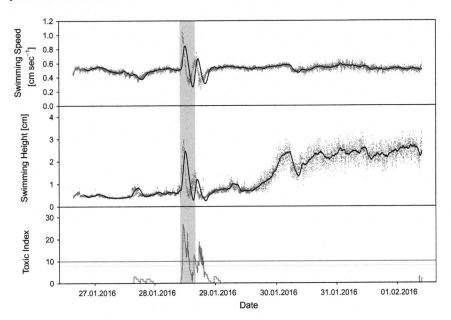

Fig. 5.21 Progression of the parameters swimming speed (upper), swimming height (middle) and toxic index (lower) during the Metoprolol experiment. The red area indicates the exposure phase with Metoprolol. Grey dots are the data points, the solid black line is based on the moving averages of 135 data points according to the bbe algorithm. The calculated toxic index over all parameters is shown as a grey line in the lowest plot with the pre-alarm (yellow) and alarm (red) thresholds as horizontal lines

Whereas the FluoSens was able to quantify the exposure with MET, the *Daphnia*-Toximeter was again able to detect the toxicity of the occurring exposure, i.e. to qualify the exposure. In contrast to PCP the reactions of the daphnids occurred faster and were more similar between individuals. The swimming speed and height increased markedly about 30 min after the start of exposure (Fig. 5.21). Further parameters like average distance or fractal dimension showed alterations at the same time (data not shown). The pre-alarm and alarm thresholds of the toxic index were achieved both after about 60 min. It should be noted that the exposure with MET did not result in any mortality in chamber 2 and only to a slight increase of 10% in chamber 1. Hence, this exposure scenario is a good example for sublethal exposure.

Summary of exposure experiments

Both experiments clearly revealed that an exposure to toxic compounds can be qualified and to a certain extent quantified using the *Daphnia*-Toximeter and the FluoSens from bbe. The toxic index as a measure for the toxicity of the sample water showed a distinct increase in both experiments. The PCP experiment represented a worst case scenario with strongly increased mortality, whereas the MET experiment showed only sub lethal effects on daphnids, although lethal effects were expected due to the high concentration.

Comparing both experiments revealed a much faster response of the *Daphnia*-Toximeter to MET compared to PCP, although the mortality was much higher under PCP exposure. This is an interesting effect and seems to be correlated with the physico-chemical parameters of both substances. The higher water solubility and the reduced adsorptive capabilities of MET led to a more equal distribution of the test substance within the system and accordingly to a faster exposure of daphnids resulting in a quicker response. This is also visible in the SAC255 data from the FluoSens, which showed a much more uniform response to MET compared to PCP.

5.6.3 *Biomonitoring in Germany - Field Study with the Daphnia-Toximeter at the Wastewater Treatment Plant Kreischa (Germany)*

From March until end of October 2016 a field study at the waste water treatment plant Kreischa (WWTP Kreischa) near Dresden (Germany) was performed to gain experience with the *Daphnia*-Toximeter and the FluoSens under field conditions. The monitoring station was part of the monitoring system "Lockwitzbach", which has been established by the Institute of Hydrobiology and the Institute of Urban Waters of the TU Dresden and is composed of different-sized monitoring stations. The major aim of the study was to create a prototype monitoring station to optimise the connectivity among the different instruments and probes as well as to the project data base AL.VIS. Moreover, the station was used as a showcase for conference and workshop presentations, international guests and students to demonstrate the general

Fig. 5.22 Monitoring station Kreischa. Left – View into the monitoring trailer showing Dr. Marcus Rybicki explaining the *Daphnia*-Toximeter to staff of the WWTP Kreischa. Right – Dr. Dirk Jungmann at the effluent of the WWTP Kreischa explaining the double circular system of the monitoring station to Prof. Wang Hong-Zhu and Prof. Wang Haijun from the CAS Institute of Hydrobiology in Wuhan (Source: M. Rybicki)

functioning of a small monitoring station and the prototype monitoring network "Lockwitzbach".

The location at the WWTP Kreischa was chosen to investigate the complex composition of the WWTP effluent with the *Daphnia*-Toximeter. The City of Kreischa harbours a health clinic, which produces a significant part of the waste water. Hence, the waste water was expected to contain pharmaceuticals that are partly or not degradable in the treatment plant (Gurke et al. 2015) and may cause toxic effects to daphnids in the *Daphnia*-Toximeter.

The housing of the monitoring station was provided by a trailer. All instruments and devices as well as supporting material were stored in the trailer. Power supply was provided by the WWTP Kreischa. The connection of the monitoring station to the project data base was set-up by our project partner AMC via mobile service. The sample water from the effluent was set-up using a simple raw water pump (GARDENA 6000, Gardena, Germany), which circulated the water from the effluent through the station back to the effluent. Unfortunately, the treatment procedures in the WWTP caused regular phases without effluent water especially during night-time (2 to 4 AM). Therefore, a double circulation system with two water pumps and a standard water barrel (150 L) as reservoir was established (Fig. 5.22). Due to the low oxygen concentrations in the WWTP effluent (≤ 4 mg L^{-1}), an aeration of the sample water in the water barrel was established using a simple aquarium air pump and sparkling stones. The station was equipped with the *Daphnia*-Toximeter to monitor the toxicity of the effluent water and with the FluoSens to roughly analyse the composition of the effluent water. Further probes had not been available at that stage of the project. The instruments were maintained regularly, including the exchange of the daphnids

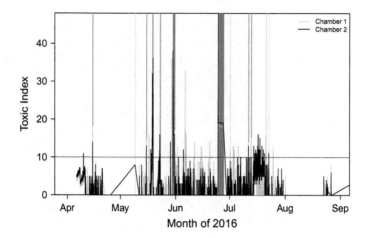

Fig. 5.23 Progression of the toxic index of the single chambers (grey and black solid lines) of the *Daphnia*-Toximeter during field study at the WWTP Kreischa based on the data send to the AL.VIS-Database (5 min data). The vertical lines represents times of pre-alarms (yellow) and alarms (red)

and cleaning of all parts, at least once a week. In case of increased mortality, fresh daphnids were introduced and necessary maintenance was performed. Feeding of the daphnids with 10 $\mu L\ h^{-1}$ bbe algae solution was stopped at the 6th of July to reduce the growth rate of the daphnids and to shift the start of the reproduction backwards.

Results and Discussion

The results of the toxic index of the *Daphnia*-Toximeter transmitted to the project data base AL.VIS are shown in Fig. 5.23. As indicated by the gaps in the time series, the measurements during the field study were interrupted twice, in April/May and August, due to damages at the *Daphnia*-Toximeter, which were caused by a defective power supply unit. However, from mid of May until end of July an uninterrupted measurement was performed.

Generally, daphnid behaviour showed a stronger variation compared to the laboratory experiments and regularly triggered pre-alarm (toxic index ≥ 8) and alarm states (toxic index ≥ 10) in the chambers of the *Daphnia*-Toximeter. However, simultaneous pre-alarms and alarms in both chambers, which triggers an instrument pre-alarm/alarm, occurred only on a few dates as indicated by the vertical yellow and red lines of Fig. 5.23. The observed alarms could be assigned to different causes. Most of the alarms were caused by technical difficulties, which affected the detection of the daphnids or their survival. For instance, the alarm at the end of May was caused by strong reproduction of the daphnids with more than 30 specimen per chamber, which led to a total malfunction of the detection algorithm and an alarm in both chambers. The longer alarm period at the end of June was also caused by technical difficulties, where the cooling of the sample water failed due to heat accumulation in the trailer. This led to sample water temperatures of 38.5 °C and the death of

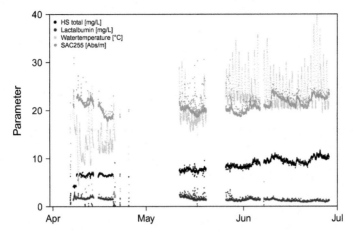

Fig. 5.24 Progression of total humic substances (HS total), peptides (Lacatalbumin), SAC255 and water temperature at the effluent of the WWTP Kreischa determined by the bbe FluosSens

all daphnids within both chambers. Apart from these technical based alarms also alarms without technical reason occurred. Especially the alarms during mid of May (15./19./22./23.05.2016) could not be assigned to technical difficulties and seemed to be correlated with toxicity. This was supported by the observed increased mortality during this period. Due to the prototypic installation of the station no further analysis of these alarm events was possible. Additional data provided by the WWTP had a low temporal resolution and revealed no further information about the causes of the observed toxicity.

The FluoSens could be used until end of June 2016 with two interruptions. Figure 5.24 shows the progression of the water temperature, the total concentration of humic substances, peptides as well as the SAC255 as a measure for the total organic pollution. As expected for waste water treatment effluents, the concentration of humic substances and peptides as well as the SAC255 were relatively stable during the entire period, although the SAC255 fluctuated slightly stronger compared to the other parameters. A relation to the observed toxicity especially during mid of May was not evident. Only the water temperature especially at the end of June correlated with the observed alarms of the *Daphnia*-Toximeter, which was caused by the malfunction of the sample water cooling as described above.

Summary

In total, the field study was successful and revealed important information about the biofouling of both instruments and the required maintenance effort under field condition as well as the behaviour of the instrument at low (≤ 10 °C) and high (≥ 35 °C) ambient temperatures. Furthermore, the data collection and transmission were established and optimised.

The *Daphnia*-Toximeter generally operated stably, but the regularly occurring alarms in single chambers underlined the necessity of a two-chamber version for a valid biomonitoring. This solution makes the system more robust and allows a replication of the measurement. Originally, more alarms were expected due to the increased exposure to pharmaceuticals in the WWTP effluent. The low number of alarms illustrates the generally suitable treatment performance of the WWTP and the applicability of the *Daphnia*-Toximeter under field conditions. In contrast to the original expectations, the daphnids showed even a very fast growth in the *Daphnia*-Toximeter during exposure to effluent water, which finally forced us to stop additional feeding with algal solutions in order to prevent a too early reproduction.

A rough quantification of effluent water composition using the FluoSens was possible, although no correlation between the determined carbon groups or the SAC255 with the observed toxicity was found. Nevertheless, the FluoSens enabled a basic quantification of important parameters, which can be used to exclude and refine possible causes for the observed toxicity.

5.6.4 Biomonitoring in China - Establishment and Operation of the Daphnia-Toximeter at Bao'an Lake

Biomonitoring with the *Daphnia*-Toximeter in China was performed in close cooperation with the group of Ecology and Taxonomy of Benthos at the Institute of Hydrobiology, Chinese Academy of Sciences (IHB-CAS) in Wuhan. As a suitable location for the biomonitoring, the pond mesocosm facility of IHB-CAS east to Lake Bao'an (Sect. 5.3) was selected. At this facility, different ponds with unique chemical characteristics, especially regarding nitrate and ammonium concentration from recent experiments, could be used for the monitoring. The transfer of the *Daphnia*-Toximeter and all additional equipment to Wuhan was performed end of May 2017. The monitoring station with instruments from Germany and additional probes for nitrate and ammonium (VARION PLUS 700IQ, WTW) from the CAS Institute of Hydrobiology was then established in July 2017 after constructional upgrades of the housing (Fig. 5.25).

The monitoring in the field station was much more difficult compared to the field study in Germany. The major challenge was the power supply of the station. Although an uninterrupted power supply (UPS) was provided from the Chinese partners and available in the station, its capacity was limited to ca. 12 h. Hence, longer interruptions of the power, which occurred regularly during summer, forced several stops of the monitoring. While the monitoring of physico-chemical parameters was relatively robust and could be applied at most times, the biomonitoring was strongly affected by the unstable power supply. The *Daphnia*-Toximeter requires an external water supply, provided by a pump in the respective pond which could not be connected to the UPS. Furthermore, an air condition, which could not be connected to the UPS neither, was required for operation, as water cooling capabilities of the *Daphnia*-Toximeter were impaired at air temperatures >35 °C. Consequently, it was not possible to operate the instrument during times of unstable power supply. Finally, a damage of the *Daphnia*-Toximeter caused an additional longer monitoring break of the biomonitoring in autumn 2017.

Fig. 5.25 Impressions from the field station at the Bao'an lake near Wuhan. Upper left and right: housing of the monitoring station before (June 2017) and after (November 2017) constructional upgrades. Lower left: view into the compiled monitoring station with *Daphnia*-Toximeter, IQ SensorNet controler, uninterruptible power supply and a backup data collector. Lower right: swimming basket with pump for raw water withdrawal and IQ SensorNet probes in an experimental pond of the field station (Source: M. Rybicki)

First Results

While the monitoring in the field station is still ongoing and under steady optimisation, first results can be shown. Figure 5.26 shows the results from the physico-chemical monitoring at the field station until January 2018. The recorded data can be distinguished into two separate phases. The first phase includes the summer and autumn season until October, whereas the second phase started at the beginning of November after major maintenance and inspection of the monitoring equipment.

In total, the physico-chemical monitoring could be performed relatively continuous except for a gap over a few weeks during October, which is quite good considering the unstable power supply especially in August 2017. The obvious leaps of nitrate and ammonium during the first phase resulted from relocations of the SensorNet-probes into experimental ponds with different chemical characteristics. Very noticeable is the increasing diurnal amplitude of the oxygen concentration, especially during July, achieving a maximum concentration of 20 mg L^{-1} at 37.2 °C recorded at the 28th of July at 3:29 PM, which corresponds to an oxygen saturation of ca. 300%. This extreme high value was a result of a cluster of filamentous green algae surrounding

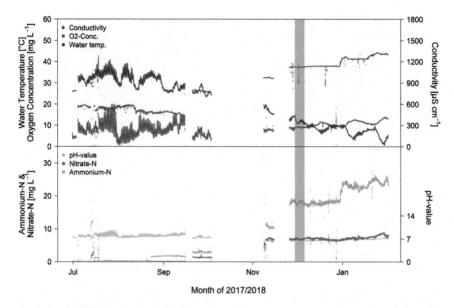

Fig. 5.26 Physico-chemical data from the monitoring at the Bao'an Lake near Wuhan. The blue-green area indicates the time of stable biomonitoring measurement in December 2017 as discussed in the text

the probes and, hence, underlines the importance of regularly maintenance (QA) in terms of a good quality management, as discussed in Sect. 5.5.

The second phase started in November 2017 with major maintenance and validation measures in the monitoring station. Validation and recalibration of the probes were performed at the 9th of November and are clearly indicated by the increase of nitrate and ammonia after recalibration. Subsequent data acquisition until January 2018 was stable for all parameters, although the water temperature showed a season-dependent decrease. The visible increase of ammonium-N was caused by intended addition of ammonium chloride to the experimental ponds. Although the quality assurance was optimised in the second phase, the data still show single outliers, e.g. in the conductivity at the 2nd of December. These outliers are normal to occur in long-term high-frequency measurements and are good examples of the necessity for data post-processing within the data quality control during online monitoring.

As explained above, the *Daphnia*-Toximeter was only occasionally in operation, due to the unstable power supply and technical difficulties. Exemplary results from a measurement in December 2017 are presented in Fig. 5.27. The results show a generally stable measurement with occasional increases of the toxic index, which stayed below the pre-alarm threshold. Average swimming speed was very stable indicating good conditions of the daphnids in the observed chamber of the *Daphnia*-Toximeter. The average swimming height was relatively stable, too, and showed no large increases, which would be a common sign of stress or intoxication. Data analysis showed that the sporadically observed increases of the toxic index were primarily

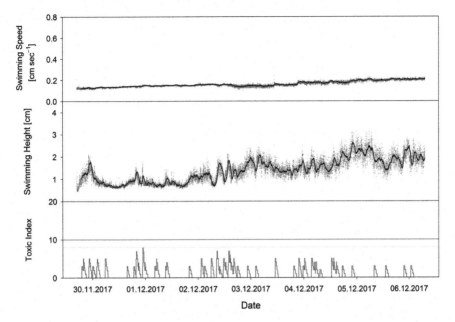

Fig. 5.27 Progression of the parameters swimming speed (upper), swimming height (middle) and toxic index (lower) during application of the *Daphnia*-Toximeter to an ammonium enriched pond of the mesocosm facility at Bao'an Lake. Grey dots are the data points, the solid black line is based on the moving averages of 135 data points according the bbe algorithm. The calculated toxic index over all parameters is shown as a grey line in the lowest plot with the pre-alarm (yellow) and alarm (red) thresholds as horizontal lines

caused by the average swimming height. The natural variation of this parameter is relatively large, even under standardized laboratory conditions (see e.g. Sect. 5.6.2) and explains the observed variation. Considering that all other determined behavioural parameters of the *Daphnia*-Toximeter were inconspicuous, the measurement can be rated as stable without toxic events.

This result is quite interesting considering the determined ammonium concentration in the sample water. The VARION 700 IQ probe determined an average concentration of 17.8 ± 0.4 mg L^{-1} ammonium-N in the pond. Using the normalisation method described in US-EPA (2013) [normalised to pH 7 and 20 °C] led to a total ammonia-N value (TAN) of 6.5 ± 0.7 mg L^{-1} TAN. Considering the reported EC$_{50}$ for *Daphnia magna*, which ranges between 37 and 419 mg L^{-1} TAN (US-EPA 2013), revealed a factor range of 5.6–64.2 to the reported effect values. Hence, behavioural anomalies of the daphnids seemed to be quite likely. However, no signs of toxicity or increased stress were found during the measurement. One reason for the missing toxicity may be the better conditions for daphnids in complex natural water from the pond mesocosms compared to standardized laboratory media, e.g. a higher availability and a more diverse composition of bacteria and algae. This may increase the physiological fitness of daphnids and, hence, their tolerance to environmental stressors. Similar effects have been observed by Wang et al. (2017) in experiments with different fish species in these mesocosms. Additionally, the physico-chemical characteristics during December should be considered. The prevailing pH values and water temperatures finally led to a relatively low proportion of ammonia and a high proportion of ammonium. Calculated concentration of un-ionized ammonia in the pond reached only 0.010 ± 0.001 mg L^{-1}, which is far away from the calculated normalised ammonia concentrations (TAN). Hence, the ambient conditions in the ponds seemed to prevent toxic effects on daphnids.

Summary

The difficulties of the biomonitoring in China have been solved and were mainly caused by the unstable power supply in the suburban areas of Wuhan at the Bao'an lake. Fostered by the fruitful joint Sino-German cooperation, a fully functional biomonitoring station could be established. Some work is still pending and currently being implemented, but first successful measuring campaigns have been performed. Monitoring of the physico-chemical parameters worked very well, although quality assurance measures are still in the process of adaptation and optimisation according to the local situation. The *Daphnia*-Toximeter was successfully installed and first measurements could be performed in the autumn/winter 2017. Apart from the temporarily unstable power supply, which is a major challenge for continuous monitoring, the local quality assurance for the biomonitoring, involving the process of culturing daphnids, as well as the quality assurance measures at the instrument itself, is still being optimised. However, the first results gained and presented here are very promising and further experiments, especially on the impacts of ammonia on the daphnids, are already scheduled for 2018. Finally, the performed experiments and measurements already have risen the interest of local stakeholders on biomonitoring.

5.7 Three-Dimensional Modelling as a Supporting Tool for Lake Management

Marieke Frassl, Weiping Hu, Karsten Rinke

Chao Lake is a very shallow, but, as measured by surface area, large lake. This large surface area to depth ratio renders Chao Lake highly sensitive towards wind forcing. Given its relatively short residence time of about half a year, it is furthermore strongly influenced by its inflows. This is, for example, reflected by higher nutrient concentrations in the western part of the lake arising from highly polluted riverine inputs at the western shore (Zan et al. 2011). The reaction of the lake to external forcing (e.g. wind) and the internal transport and distribution of pollutants or cyanobacteria within the lake are important dynamic aspects relevant for lake managers. We therefore included the application of a three-dimensional lake model in our environmental information framework. We employed the model GETM as outlined in Sect. 5.4.4.

The tasks associated with the application of the three-dimensional lake model were as follows:

1. Generate a consistent input data set for the model out of scarce on-site measurements.
2. Perform own measurements and establish a suitable data set that allows a test application of the model.
3. Simulate Chao Lake and show that the model is able to capture relevant temporal and spatial dynamics.
4. Use the model to estimate the potential for wind-induced resuspension of bottom sediments.

The application of three-dimensional models is a demanding task, not only with respect to computational requirements, but also regarding the availability and quality of input data. In our case, time-series of meteorological input variables (air temperature, air pressure, short-wave and long-wave radiation, wind speed and direction, and humidity) and hydrological discharge data (inflows and outflows) were required. In many cases, the availability of these data is limited as it was in our test case on Chao Lake. We, therefore, developed a modelling approach that minimised the input data requirements in order to allow a pragmatic, but still informative, application of the model. This included two important simplifications:

1. We left out all hydrology by setting inflow and outflow discharges to zero and excluded rain as well as evaporative loss of water from the lake, i.e. the change of volume over time was zero. Note that evaporative heat loss (cooling) was still included.
2. Instead of using meteorological measurements from local field stations (normally only accessible under high costs and effort), we used meteorological data from a global reanalysis (see Sect. 5.7.2). These data have a global coverage, are free and can be downloaded easily.

In summary, our three-dimensional model fully resolved the thermodynamics of the lake (i.e. the heat budget), but was simplified in resolving hydrodynamics in the way that inflow/outflow dynamics were not included. Advection, convection and mixing within the lake were therefore solely driven by meteorological variables and not by the water renewal. This assumption was fully valid for short-term dynamics (days to a few weeks), but incorrect over longer time-series because water imported by the inflow and delivering, e.g., nutrients or pollutants, was not taken into account. As a consequence, our approach is poor for predicting water quality dynamics (e.g. nutrient concentration and algal bloom formation) but strong for predicting temperature dynamics, transport within the lake, and resuspension potential.

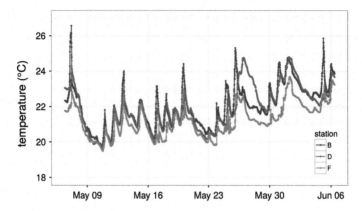

Fig. 5.28 Water temperature data from the surface measured at three different locations in Chao Lake. The positions are denoted by the station name as given in Fig. 5.10

5.7.1 Generating a Consistent Input Data Set for the Model: 3D-Monitoring of Temperature Dynamics

We were able to collect a consistent three-dimensional data set of temperature dynamics within Chao Lake through our logger deployments outlined in Sect. 5.4.5. Water temperature closely followed the seasonal course of its meteorological drivers (compare Fig. 5.13, more details are shown in Sect. 5.7.3), but also showed some horizontal differences (Fig. 5.28) that frequently reached values in the range of 2–3 centigrades.

Despite the shallow depth and intense vertical wind mixing in Chao Lake, we observed intermittent intervals of temperature gradients along the vertical axis indicating thermal stratification of the water column (Fig. 5.29). This stratification often persisted only over a day and was eroded completely by convective cooling overnight. Sometimes, during conditions of strong radiative forcing and low wind speed, magnitudes of about 5 centigrades were reached. Such temperature differences prevented vertical mixing and therefore could go along with decreasing oxygen concentrations in the deep waters (e.g. around April 16 in Fig. 5.29). The rather fast drop in oxygen concentrations under conditions of reduced vertical mixing needed to be contributed to the high organic pollution of Chao Lake leading to high biological oxygen demands. These measurements illustrated that stratification events could be an important factor for short-term oxygen dynamics. An appropriate simulation of the stratification patterns found in the water temperature data constituted a good test case for critically evaluating the ability of the three-dimensional model to simulate the hydrodynamic characteristics of large shallow lakes.

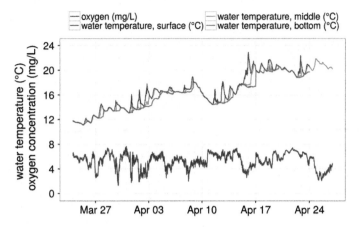

Fig. 5.29 Water temperature at the surface and above the bottom as well as the oxygen concentration above the bottom measured at station B (see Fig. 5.10)

5.7.2 Applicability of Re-analysis Data as Meteorological Input Data

As described above, the time-series of meteorological variables required by the model are very often difficult to receive. They are either only made available at high financial costs (in China required data often have to be bought), but the larger hurdle is a lack of availability, e.g. due to a general low spatial coverage with meteorological surface stations or a lack of data required by the model. The latter often accounts for variables like cloud cover or short-wave and long-wave radiation as most meteorological stations are not equipped with the required sensors. We therefore obtained the meteorological input data from the global reanalysis ERA-Interim (Dee et al. 2011), which is being developed and maintained by the European Centre for Medium-Range Weather Forecasts (ECMWF). The required data can be downloaded from the public web interface available at www.ecmwf.int, where we downloaded meteorological data for the variables air temperature (K), dewpoint temperature (K), 10 m u and v wind speed components (ms^{-1}), total cloud cover (-) and mean sea level pressure (Pa). The data were downloaded at 6-h time steps from the analysis (00:00, 06:00, 12:00 and 18:00) and 6-h time steps as forecast values (at 03:00, 09:00, 15:00 and 21:00), resulting in a temporal resolution of 3 hours. For the simulation, meteorological data were linearly interpolated to hourly time steps. The spatial resolution of the downloaded data was 0.125° and we used data from the centre of the lake (31.5°N and 117.5°E).

We compared the retrieved reanalysis data with meteorological data measured at a field station close to Chao Lake in order to assess the deviations between measurements and reanalysis. For some meteorological variables, the reanalysis data closely followed observations (e.g. air temperature, see Fig. 5.30), but variables stronger influenced by local conditions showed higher deviations (e.g. wind direction and

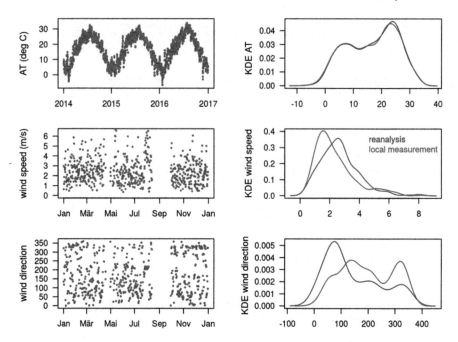

Fig. 5.30 Comparison of measured meteorological data (red) and meteorological data from a reanalysis model (blue) with respective kernel density estimates (KDE). Top: air temperature (°C), middle: wind speed (m/s), bottom: wind direction (°)

wind speed, see Fig. 5.30). Overall, we were positively surprised by the good agreement of reanalysis data with the local conditions. Accordingly, we assumed that they were suitable as input data for a three-dimensional lake model.

5.7.3 Three-Dimensional Hydrodynamic Simulation of Chao Lake

Simulations with GETM showed the highly dynamic character of Chao Lake. Seasonal forcing of water temperatures was strong and the lake reacted fast to meteorological conditions due to its shallowness and low heat storage. Maximum temperatures reached more than 30 °C in July and minimum temperatures went below 5 °C in January. This high temperature range reflected the climatic conditions of Chao Lake being located at the northern edge of the subtropical zone. Simulated water temperatures from GETM complied very well with the temperature recordings from the deployed temperature loggers (see Sects. 5.4.5 and 5.7.1). This held true for surface water temperature (Fig. 5.31) as well as for bottom water temperatures (Fig. 5.32). In conclusion, the good agreement between observation and simulation did not only

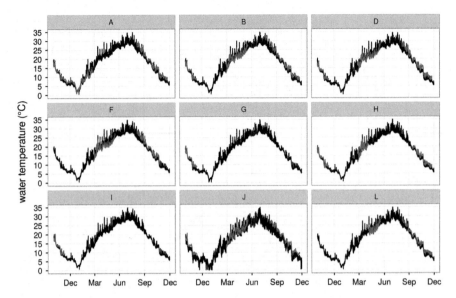

Fig. 5.31 Measured (red) and simulated (black) surface water temperatures in Chao Lake at different sampling locations (see Fig. 5.10) over the simulation period (Nov 2015–Dec 2016). Data gaps in measured water temperatures are due to lost temperature loggers

indicate that GETM was able to simulate the thermodynamics of the lake with high accuracy. It also illustrated that the used reanalysis data provided a reasonably good data basis for simulating temperature dynamics of the water body. This was an important outcome as this approach was fully transferable to other lakes in China or other places on the globe (see also Frassl et al. 2018).

Simulations with GETM also revealed horizontal heterogeneities in Chao Lake. Water temperatures, for example, varied up to 5 centigrades horizontally at the lake surface (Fig. 5.33) during specific times of the year. These spatial differences were induced by wind effects, e.g. by upwelling of colder water at the upwind side. It should be emphasised that inflows as well as heterogeneous meteorological forcing was not responsible for these differences in the simulations because inflows were turned off and meteorological forcing was provided as a horizontally homogeneous field, i.e. meteorological conditions at the western side were always the same as at the eastern side in the simulation. This was certainly not always true in reality given the large size of Chao Lake (approximately 50 km from east to west). Moreover, most inflows in Chao Lake were located in the western part and the only outflow was in the eastern part. The magnitude of horizontal differences in reality might therefore be even larger than in the simulation. Nevertheless, the model provided a quantitative impression of horizontal heterogeneities.

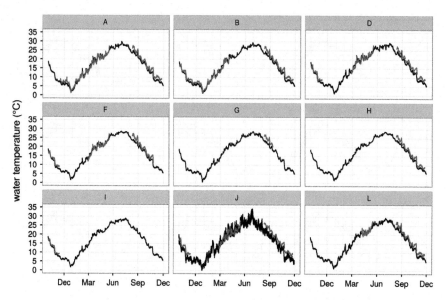

Fig. 5.32 Measured (red) and simulated (black) bottom water temperatures in Chao Lake at different sampling locations (see Fig. 5.10) over the simulation period (Nov 2015–Dec 2016). Data gaps in measured water temperatures are due to lost temperature loggers

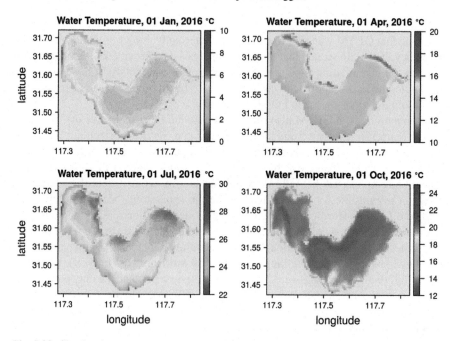

Fig. 5.33 Simulated water temperature at the surface of Chao Lake during different times of the year 2016

5.7.4 Resuspension Events

Wind-induced resuspension of bottom sediments can frequently occur in shallow lakes (Hamilton and Mitchell 1996; Hawley and Lesht 1992). Wind has an immediate influence on the velocities at the sediment-water-interface in these lakes either by wind-induced surface waves or by inducing large scale currents. Resuspension events are very important for the water quality dynamics as they usually go along with significant nutrient release from the sediments into the open water body (Hamilton 1997; Li et al. 2017b), potentially driving eutrophication. In lakes with high pollution by heavy metals or other toxicants, the sediments often act as a deposition site for these pollutants and resuspension events can release pollutants into the water (Williamson et al. 1996).

While the prediction of resuspension events is strongly site-specific due to the site-specific sediment properties (e.g. mud, silt, sand, ...), hydrodynamic models can be used to provide physical information on the forces at play at the sediment-water interface. The critical quantity for resuspension of sediments is the bottom shear stress. A shear stress describes the force that acts parallel to a surface, in our

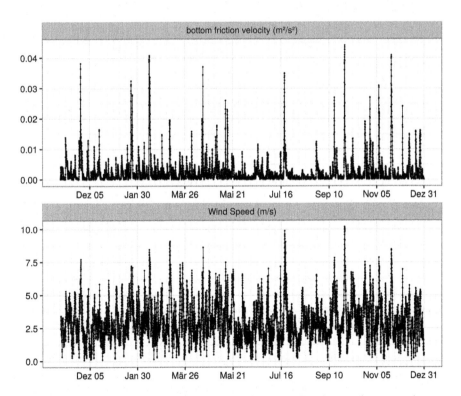

Fig. 5.34 Comparison of bottom shear stress and wind stress. Top: bottom shear stress, bottom: wind speed

case the force arising from water currents over the sediment surface. This bottom shear stress scales linearly with the square of the bottom friction velocity ($m^2\ s^{-2}$). Both quantities are linked by the drag coefficient, which depends slightly on local surface characteristics. We used GETM to calculate bottom friction velocities over the simulation period and compared these values with wind velocities at Chao Lake (Fig. 5.34). The time-series of bottom friction velocities appeared spiky and high values were only reached during short periods when wind was sufficiently strong.

We refrained from translating numerically evaluated bottom friction velocities into resuspension rates because the physical properties of the lake sediments in Chao Lake were unknown and further in-situ studies would be required to build a quantitative connection for sensible predictions. However, the model output showed that only events with high wind speeds were able to induce high friction velocities. A critical wind speed for that is approximately around 6 ms^{-1} (compare Fig. 5.34). More detailed studies on wind-induced sediment resuspension were undertaken in Lake Taihu (Li et al. 2017b), which has similar depth and surface area and is only a few hundred kilometres away from Chao Lake. Interestingly, Li et al. (2017b) found sediment resuspension in Lake Taihu to increase significantly as soon as wind speeds surpass 6 ms^{-1}; a fact that fully complies to our model-based assessment of bottom friction velocity in Chao Lake.

5.8 Synopsis

Marcus Rybicki, Bertram Boehrer, Karsten Rinke, Marieke Frassl

We present good progress in practical aspects of water quality monitoring in lakes. This spans from in situ monitoring, e.g. by buoy-based systems or deployed loggers, to on-site biomonitoring. While the former is the method of choice at the ecosystem scale, the latter is most suitable for water infrastructures like water abstraction or drinking water production. We ran these sensitive infrastructures remotely from their operators and established reliable data transfer and visualisation infrastructures that made them readily applicable in water management.

The successful implementation of biomonitoring in China is a great achievement. Water quality assessment by authorities in China is currently strongly focused on chemical analysis, while integrative measures like biomonitoring, also in the classical sense of bioindication, are primarily used for scientific purposes. However, the complex pollution characteristic of Chinese surface waters require an integrative assessment of the water quality. Dynamic biomonitors like the *Daphnia*-Toximeters were successfully deployed in European surface waters since the 1970s, permanently proving their applicability in water quality assessment. The successful implementation of the *Daphnia*-Toximeter already attracted attention of local stakeholders and hopefully will foster biomonitoring also beyond science. The chance to directly visualise toxicity of a water body is thereby a powerful tool to proliferate thinking about environmental pollution in authorities and in the public, which is a prerequisite for a sustainable lake management.

Furthermore, we managed to implement field investigation and numerical modelling as supplementary tool. Models were used in predictive mode referring to measured data and hence provided lake-wide, i.e. spatially resolved, information beyond what could be covered by field deployment. A direct application of models in lake management, however, requires more efforts. The required input data need to be provided in a fast and easy-to-use mode, possibly even by online-transfer from the authorities responsible for hydrological and meteorological measurements. To operate a lake model in online-mode is a tremendous effort that deserves a project on its own. It is, however, a powerful tool, e.g. for the prediction of pollutant transport or the analysis of alternative management strategies.

The most crucial point in science-based lake management, however, is a close interaction between scientists and managers. This is the most important and most difficult aspect. Funding of scientific projects helps to push the development and implementation of powerful scientific tools in monitoring and modelling. This goes faster than the build-up of administrative structures that make use of these developments. For sensitive, but stressed, water bodies, e.g. lakes or reservoirs in heavily populated areas used for drinking water provision, it pays off to invest efforts in model- and monitoring-aided water quality management.

Acknowledgements We are grateful to the German Academic Exchange Service for the financial support provided in the project "Water systems of the Yangtze River Basin". We thank Burkhard Kuehn for sustained technical support and for preparing the deployment of the buoy. The Collaborative Research Centre TRR 181 on Energy Transfers in Atmosphere and Ocean, funded by the German Research Foundation, is thanked for the financial support of Knut Klingbeil. Chen Yun is thanked for technical support of the buoy deployment in Chao Lake and finally Wu Fengfu is thanked for the technical support at the mesocosm facility at the Bao'an Lake.

References

Becherer J, and Umlauf L. Boundary mixing in lakes: 1. Modeling the effect of shear-induced convection. J. Geophys. Res.: Oceans **116**(10), (2011). https://doi.org/10.1029/2011JC007119

Bruggeman J, and Bolding K. A general framework for aquatic biogeochemical models. Environ. Model. Softw. **61**, 249–265 (2014). https://doi.org/10.1016/j.envsoft.2014.04.002

Burchard H, and Bolding K. GETM – a General Estuarine Transport Model. Scientific Documentation. Technical Report EUR 20253 EN, European Commission (2002a)

Burchard H, Bolding K, and Villarreal MR. GOTM – a General Ocean Turbulence Model. Theory, implementation and test cases. Technical Report EUR 18745 EN, European Commission (1999)

Burchard H, Bolding K, and Villarreal MR. Three-dimensional modelling of estuarine turbidity maxima in a tidal estuary. Ocean Dyn. **54**, 250–265 (2004). https://doi.org/10.1007/s10236-003-0073-4

Campbell JL, Rustad LE, Porter JH, Taylor JR, Dereszynski EW, Shanley JB, Gries C, Henshaw DL, Martin ME, Sheldon WM, and Boose ER. Quantity is nothing without quality: automated QA/QC for streaming environmental sensor data. BioScience **63**(7), 574–585 (2013). https://doi.org/10.1525/bio.2013.63.7.10

Dee DP, Uppala SM, Simmons AJ, Berrisford P, Poli P, Kobayashi S, Andrae U, Balmaseda MA, Balsamo G, and Bauer P et al. The ERA-Interim reanalysis: configuration and performance of the data assimilation system. Q. J. R. Meteorol. Soc. **137**(656), 553–597 (2011). ISSN 1477-870X. https://doi.org/10.1002/qj.828

Frassl MA, Boehrer B, Holtermann PL, Hu W, Klingbeil K, Peng Z, Zhu J, and Rinke K. Opportunities and limits of using meteorological reanalysis data for simulating seasonal to sub-daily water temperature dynamics in a large shallow lake. Water **10**(5), 594 (2018). https://doi.org/10.3390/w10050594

Green U, Kremer JH, Zillmer M, and Moldaenke C. Detection of chemical threat agents in drinking water by an early warning real-time biomonitor. Environ. Toxicol. **18**(6), 368–374 (2003)

Gurke R, Rößler M, Marx C, Diamond S, Schubert S, Oertel R, and Fauler J. Occurrence and removal of frequently prescribed pharmaceuticals and corresponding metabolites in wastewater of a sewage treatment plant. Sci. Total Environ. **532**, 762–770 (2015). ISSN 0048-9697. https://doi.org/10.1016/j.scitotenv.2015.06.067

Hamilton DP. Wave-induced shear stresses, plant nutrients and chlorophyll in seven shallow lakes. Freshw. Biol. **38**, 159–168 (1997)

Hamilton DP, and Mitchell SF. An empirical model for sediment resuspension in shallow lakes. Hydrobiologia **317**(3), 209–220 (1996). ISSN 1573-5117. https://doi.org/10.1007/BF00036471

Hawley N, and Lesht BM. Sediment resuspension in lake St. Clair. Limnol. Oceanogr. **37**(8), 1720–1737 (1992). ISSN 1939-5590. https://doi.org/10.4319/lo.1992.37.8.1720

Hofmeister R, Burchard H, and Beckers J-M. Non-uniform adaptive vertical grids for 3D numerical ocean models. Ocean Modell. **33**, 70–86 (2010). https://doi.org/10.1016/j.ocemod.2009.12.003

Ihle T, Jähnichen S, and Benndorf J. Wax and wane of Microcystis (Cyanophyceae) and microcystins in lake sediments: a case study in Quitzdorf Reservoir (Germany). J. Phycol. **41**(3), 479–488 (2005). https://doi.org/10.1111/j.1529-8817.2005.00071.x

Klingbeil K, and Burchard H. Implementation of a direct nonhydrostatic pressure gradient discretisation into a layered ocean model. Ocean Modell. **65**, 64–77 (2013). https://doi.org/10.1016/j.ocemod.2013.02.002

Klingbeil K, Mohammadi-Aragh M, Gräwe U, and Burchard H. Quantification of spurious dissipation and mixing - discrete variance decay in a finite-volume framework. Ocean Modell. **81**, 49–64 (2014). https://doi.org/10.1016/j.ocemod.2014.06.001

Klingbeil K, Lemarié F, Debreu L, and Burchard H. The numerics of hydrostatic structured-grid coastal ocean models: state of the art and future perspectives. Ocean Modell. **125**, 80–105 (2018). ISSN 1463-5003. https://doi.org/10.1016/j.ocemod.2018.01.007

Klingbeil K, Trolle D, Schüler L, Bruggeman J, and Bolding K. A new lake model with state-of-the-art turbulence closure. Environ. Modell. Softw. (2018b in prep)

Klüttgen B, Dülmer U, Engels M, and Ratte HT. ADaM, an artificial freshwater for the culture of zooplankton. Water Res. **28**(3), 743–746 (1994)

Kong X, He Q, Yang B, He W, Xu F, Janssen ABG, Kuiper JJ, van Gerven LPA, Qin N, Jiang Y, Liu W, Yang C, Bai Z, Zhang M, Kong F, Janse JH, and Mooij WM. Hydrological regulation drives regime shifts: evidence from paleolimnology and ecosystem modeling of a large shallow Chinese lake. Glob Change Biol. **23**(2), 737–754 (2017). ISSN 1365-2486. https://doi.org/10.1111/gcb.13416

Lechelt M, Blohm W, Kirschneit B, Pfeiffer M, Gresens E, Liley J, Holz R, Lüring C, and Moldaenke C. Monitoring of surface water by ultrasensitive Daphnia toximeter. Environ. Toxicol. **15**(5), 390–400 (2000)

Li Y, Tang C, Wang J, Acharya K, Du W, Gao X, Luo L, Li H, Dai S, Mercy J, Yu Z, and Pan B. Effect of wave-current interactions on sediment resuspension in large shallow Lake Taihu, China. Environ. Sci. Pollut. Res. **24**(4), 4029–4039 (2017b). ISSN 1614-7499. https://doi.org/10.1007/s11356-016-8165-0

Moghimi S, Klingbeil K, Gräwe U, and Burchard H. A direct comparison of a depth-dependent Radiation stress formulation and a Vortex force formulation within a three-dimensional coastal ocean model. Ocean Modell. **70**, 132–144 (2013). https://doi.org/10.1016/j.ocemod.2012.10.002

Ouyang H-L, He W, Qin N, Kong X-Z, Liu W-X, He Q-S, Wang Q-M, Jiang Y-J, Yang C, Yang B, and Xu F-L. Levels, temporal-spatial variations, and sources of organochlorine pesticides in ambient air of Lake Chaohu, China. Sci. World J. (2012). ISSN 1537-744X. https://doi.org/10.1100/2012/504576

Porter JH, Hanson PC, and Lin C-C. Staying afloat in the sensor data deluge. Trends Ecol. Evol. **27**(2), 121–129 (2012). https://doi.org/10.1016/j.tree.2011.11.009

Umlauf L, and Burchard H. Second-order turbulence closure models for geophysical boundary layers. A review of recent work. Cont. Shelf Res. **25**, 795–827 (2005). https://doi.org/10.1016/j.csr.2004

Umlauf L, and Lemmin U. Interbasin exchange and mixing in the hypolimnion of a large lake: the role of long internal waves. Limnol. Oceanogr. **50**, 1601–1611 (2005). https://doi.org/10.4319/lo.2005.50.5.1601

US-EPA. Aquatic life ambient water quality criteria for ammonia - freshwater, Technical report, United States Environmental Protection Agency - Office of Water & Office of Science and Technology (Washington, DC, USA, 2013), p. 2013

Wagner M, Schmidt W, Imhof L, Grübel A, Jähn C, Georgi D, and Petzoldt H. Characterization and quantification of humic substances 2D-Fluorescence by usage of extended size exclusion chromatography. Water Res. **93**, 98–109 (2016)

Wang H-J, Wang H-Z, Liang X-M, and Wu S-K. Total phosphorus thresholds for regime shifts are nearly equal in subtropical and temperate shallow lakes with moderate depths and areas. Freshw. Biol. **59**(8), 1659–1671 (2014a). ISSN 1365-2427. https://doi.org/10.1111/fwb.12372

Wang H-J, Xiao X-C, Wang H-Z, Li Y, Yu Q, Liang X-M, Feng W-S, Shao J-C, Rybicki M, Jungmann D, and Jeppesen E. Effects of high ammonia concentrations on three cyprinid fish: acute and whole-ecosystem chronic tests. Sci. Total Environ. **598**, 900–909 (2017). ISSN 0048-9697. https://doi.org/10.1016/j.scitotenv.2017.04.070

Watson SB, Jüttner F, and Köster O. Daphnia behavioural responses to taste and odour compounds: ecological significance and application as an inline treatment plant monitoring tool. Water Sci. Technol. **55**(5), 23 (2007)

Williamson RB, Dam LFV, Bell RG, Green MO, and Kim JP. Heavy metal and suspended sediment fluxes from a contaminated, intertidal inlet (Manukau Harbour, New Zealand). Mar. Pollut. Bull. **32**(11), 812–822 (1996). ISSN 0025-326X. https://doi.org/10.1016/S0025-326X(96)00044-6

Zan F, Huo S, Xi B, Li Q, Liao H, and Zhang J. Phosphorus distribution in the sediments of a shallow eutrophic lake, Lake Chaohu, China. Environ. Earth Sci. **62**(8), 1643–1653 (2011). ISSN 1866-6299. https://doi.org/10.1007/s12665-010-0649-5

Chapter 6
WP-D: Environmental Information System

Frank Neubert, Matthias Haase, Karsten Rink and Olaf Kolditz

6.1 Motivation

The previous chapters gave detailed insight into the collection of environmental data and the use of that data for purposes such as determining and improving water quality, dealing with extreme weather events, or the planning of waste water management systems. However, adequate visualisation techniques are required to communicate the significance of this work and the consequences of research results to stakeholders or laymen. In addition, sustainable management of water resources requires well-engineered software solutions that can be operated by regional authorities and operating companies. To this end, the "Urban Catchments"-project includes the adaption and adjustment of software frameworks for the region around Chao Lake. Section 6.2 demonstrates a WebGIS approach for an environmental information system for time-series data from observation sites in Chaohu City to monitor water quality and hydro-

---

F. Neubert (✉)
AMC–Analytik & Messtechnik GmbH Chemnitz,
Heinrich-Lorenz-Straße 55, 09120 Chemnitz, Germany
e-mail: Frank.Neubert@amc-systeme.de

M. Haase
WISUTEC Umwelttechnik GmbH, Jagdschänkenstraße 50, 09117 Chemnitz, Germany
e-mail: m.haase@wisutec.de

K. Rink · O. Kolditz
Department of Environmental Informatics, Helmholtz Centre of Environmental Research–UFZ, Permoserstraße 15, 04318 Leipzig, Germany
e-mail: karsten.rink@ufz.de

O. Kolditz
Applied Environmental System Analysis,
Technische Universität Dresden, Dresden, Germany
e-mail: olaf.kolditz@ufz.de

© Springer Nature Switzerland AG 2019
A. Sachse et al. (eds.), *Chinese Water Systems*, Terrestrial Environmental Sciences,
https://doi.org/10.1007/978-3-319-97568-9_6

logical parameters. In Sect. 6.3, a virtual geographic environment is introduced to present all available data sets for the Chao Lake catchment in one unified geographic context, such that the relevance and information for each data set are adequately visualised and the interaction with other data sets is coherently illustrated. It will become obvious that such software infrastructures are becoming essential to efficiently work with the large amount of heterogenous data collected within large-scale projects.

6.2 Environmental Information System Based on Online Measurements

Frank Neubert, Matthias Haase

An important basis for the implementation of the Urban water resources management is the development and establishment of monitoring platforms for the sources of water pollution in connection with a powerful data management system. The resulting system consisting of hard- and software components at different levels is called Environmental Information System (EIS). Environmental Information Systems play an important role since they help to control, manage and reliably provide relevant data for the assessment of the existing environmental situation.

Environmental information systems are constructed in a decentralised manner and are hierarchically structured. They are modular in the target position and scalable to different dimensions (place, region, country) (Fig. 6.1). All important subsystems are covered especially when implementing environmental information systems for water balance - ground water and surface water, drinking water and waste water, households and industry. The data are associated with government information such as e.g. land registry details, properties and regulatory documents.

WISUTEC and AMC planned to implement the complete chain of an environmental information system together with all the partners based on modern technologies of the IoT (internet of things). In the "Urban Catchments"-project with Chinese partners, solutions at all levels were implemented.

Three groups of sensors are used for environmental information systems in the water and waste water area: sensors for water quality or quality parameters (temperature, conductivity, pH value, turbidity, concentration of contaminants, oxygen content, etc.), for hydrologic parameters (level, current speed, flowrate, etc.) and status parameters (camera, drop, energy supply, etc.). The selection and integration of the sensors takes place corresponding to the target position of monitoring and the local conditions (Fig. 6.2). A detailed overview on the parameters that can be measured in online operation is provided in Tables 6.1 and 6.2.

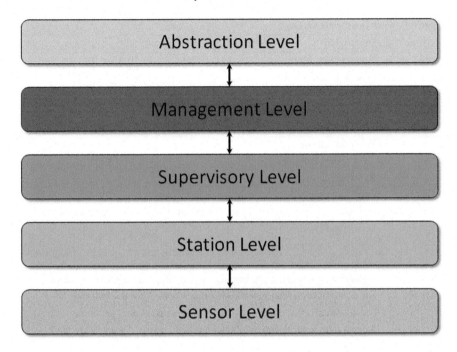

Fig. 6.1 Hierarchically structured environmental information system

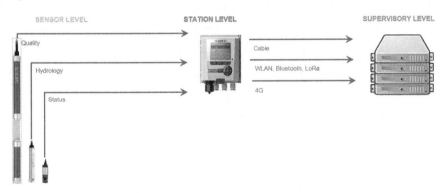

Fig. 6.2 Sensors for environmental information systems

6.2.1 Station Level

The practical implementation of the measuring stations is carried out depending on the specific local conditions with respect to measuring conditions, energy supply, communication possibilities and safety. In addition to special solutions, three standard types of stations are built: mini stations, compact stations and modular stations

Table 6.1 Water quality parameters

Parameter	Description
Temp	Water temperature
pH	Potential of hydrogen
EC	Electrical conductivity
ORP	Oxidation reduction potential
Colour	Water colour
Turbidity	Water turbidity
TSS	Total suspended solids
BOD	Biological oxygen demand
COD	Chemical oxygen demand
BTX	Aromatic hydrocarbons (mixtures of benzene, toluene and 3 xylene isomers)
TOC	Total organic carbon
DOC	Dissolved organic carbon
AOC	Assimilable organic carbon
UV254	Water quality test parameter to detect organic matter in water and wastewater
NO_3-N	Nitrate as nitrogen (concentration is being reported as nitrogen only)
NO_2-N	Nitrite as nitrogen (concentration is being reported as nitrogen only)
NH_4-N	Ammonium as nitrogen (concentration is being reported as nitrogen only)
O_3	Ozone
O_2	Oxygen
H_2S	Hydrogen sulfide
K^+	Potassium ions
F^-	Fluor ions
ClO_2	Chlorine dioxide
CO_2	Carbon dioxide
H_2O_2	Hydrogen peroxide
$C_2H_4O_3$	Peracedic acid
FC/TC	Free/total chlorine

Table 6.2 Hydrological parameters

Parameter	Description
Level	Water level
Velocity	Water stream velocity
Discharge	Water discharge
Pressure	Water pressure

which are respectively distinguished with respect to sample taking, water supply, size and energy supply (Fig. 6.3). Measuring stations are implemented in a stationary manner on the shore or bridge area, transportable as a measuring case or mobile (on boats). All communication methods are supported.

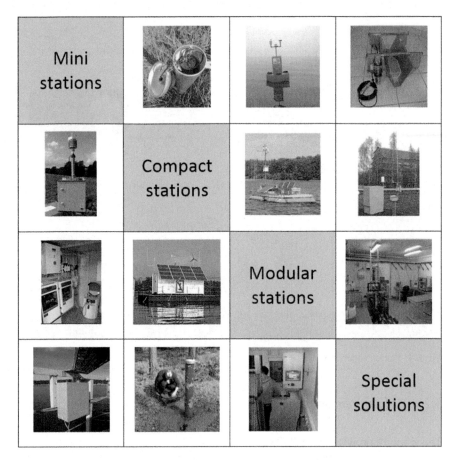

Fig. 6.3 Station level

6.2.2 Supervisory Level

A web-based solution is used as the software for the supervisory level (station supervisory system) which enables not only the monitoring of current operating statuses regarding safety, energy consumption, access rights and fault statuses, but can also execute control functions for pumps, valves and other actors. This station supervisory system (Fig. 6.4) ensures a largely maintenance-free operation and is capable of detecting faults, e.g. in pumps due to soiling in good time. Manual operation for function control on-site is also supported. All operating statuses, notifications and information on faults are recorded in a database and are available as historical data for evaluation. In the event of technical faults, notifications are actively generated.

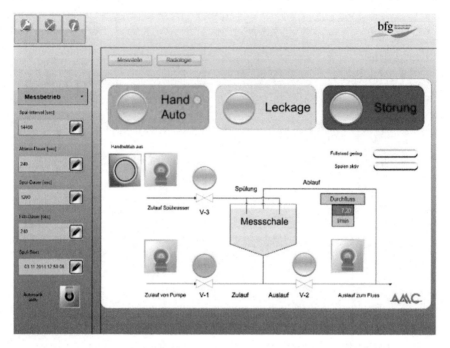

Fig. 6.4 Station supervisory system

6.2.3 Management Level

The software product, AL.VIS/Timeseries is used for the data centre of the EIS. The AL.VIS/Timeseries application is a network-based, multi-user information system for managing measurement networks and measured data in the form of time series data. It uses modern internet technologies to ensure you comprehensive access to your monitoring data. AL.VIS/Timeseries provides the user with all measured values from different areas in one system. A mobile component allows the measured data to be analysed online on smartphone or mobile device. AL.VIS/Timeseries has methods for importing data from different sources (process databases, data loggers, Excel tables), for quality assurance, research, analysis and exporting. The data-base model that is used is very flexible: additional measuring parameters or new measuring points can be easily added. The easy to use diagram function processes very long time series with thousands of measured values very quickly. AL.VIS/Timeseries is implemented in combination with web-GIS-components (Fig. 6.5).

- The software for management of monitoring data can be used in combination with software components for managing data of the river net incl. lakes
- Object-related information (cadastre) of various types
- Documents linked to objects or processes
- Survey data

6 WP-D: Environmental Information System

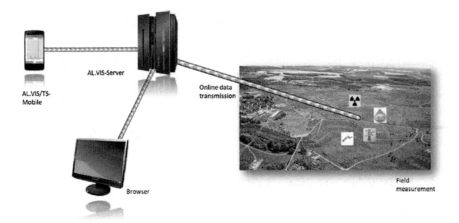

Fig. 6.5 AL.VIS/Timeseries application

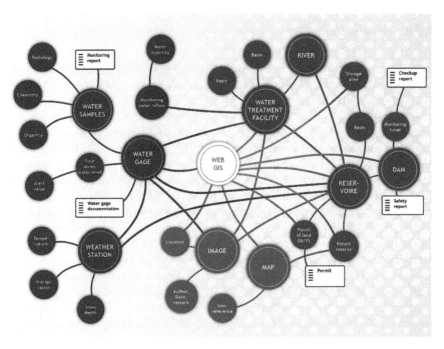

Fig. 6.6 Different information categories and their interconnections of the "Urban Catchments"-project

- Topographical, hydrological or geological maps
- Images, maps, and aerial images

Bringing together all these data under one information system allows decision makers to manage remediation processes in an effective manner. The following figure illustrates the different information categories and their interconnections (Fig. 6.6). One important goal of the "Urban Catchments"-project was to demonstrate the possibility to manage all these data and information in the EIS.

6.3 Virtual Geographic Environment

Karsten Rink

6.3.1 Motivation

In recent years, there has been a considerable progress in the development and implementation of monitoring technologies. This includes both remote sensing (e.g. the Sentinel or LANDSAT satellite missions (Fletcher 2012; Wulder et al. 2016)) and in-situ measurements (e.g. sensor networks, climate stations). In combination with advances in specialised software frameworks such as Geographical Information Systems (GIS) or widely-used services such as Google Earth, the amount of available data for a given region of interest has increased substantially. A large variety of observed data sets can be employed in the Earth system modelling domain for environmental studies in general. Of particular interest for the "Urban Catchments"-project is the analysis and management of water resources. With 40% of chinese lakes being severely polluted and 80% of its lakes suffering from eutrophication (see Sect. 1.6 for details), the exploration and understanding of complex collections of heterogeneous data sets from affected regions becomes vitally important for studies aimed at understanding possible causes and devising schemes for remediation. The water from Chao Lake, specifically, is used for fishing, irrigation and obtaining drinking water. As has been shown in Sect. 1.5, harmful algal blooms (Chen and Liu 2014; Tang et al. 2006; Wang et al. 2008) and the lake and aquifer being contaminated with fertilisers, pesticide residues and heavy metals from mining (Li et al. 2013) are of great concern.

We created a Virtual Geographic Environment (VGE) specifically for the Chao Lake catchment, utilising a Virtual Reality environment for the presentation of geoscientific and environmental data. In recent years, VGEs have been regarded as a new generation of geographic analysis tools (Chen et al. 2013, 2015b; Lin et al. 2013a, b, 2015; Lü 2011) that can be employed for a wide range of applications. This includes water pollution analysis (Walther et al. 2014), geotechnical applications (Blöcher et al. 2015) or socio-economic studies (Koch et al. 2018). Such a system allows for a combined presentation and assessment of geographic objects, such as lakes, river or sewer networks, groundwater information, etc. in combination with the results of process simulation software. Thus, VGEs can also help to predict trends and further pollution analysis (Chen et al. 2012; Lin et al. 2015) by combining observation data and modelling platforms using state-of-the art post-processing and visualisation algorithms (Helbig et al. 2014; Rink et al. 2014). Environmental Information Systems (EISs) can be embedded in VGEs to provide a realistic geographic context for a region of interest. Specifically required functionality can be added to the system, for instance so it may serve as an early warning system for drinking water supplies (see Sect. 6.2 for details).

6.3.2 Methodology

Virtual Geographic Environments provide a new and descriptive way to allow users to explore and evaluate complex collections of heterogeneous geoscientific data sets. Examples include remote sensing data, river networks, land use classes, climate data, and many more. By placing all data sets within an interactive 3D scene it becomes possible to see the correlation of data sets, the variation of certain variables in a region of interest or to highlight certain aspects of a given data set by employing established methods from visualisation sciences. In addition, numerical models and results from the simulation of (hydro-)geological processes and phenomena can also be displayed in the same context as the monitoring data.

By combining these different data sets using intuitive representations of the data, users will get a better understanding of the region of interest as well as the determining factors for phenomena or trends such as algal blooms, the accumulation of chemicals in soil and groundwater, or the effects of flooding during extreme weather events.

From the viewpoint of data management and visualisation, a number of conceptional and technical challenges have to be considered during the development of a multi-purpose Virtual Geographic Environment. For instance, each data set has to be visually represented in a way that is meaningful to a typical user. Such a representation can be either realistic or abstract. A basic example for a realistic model is shown in Fig. 6.7 where three data sets are required to create an instantly recognisable surface of the region of interest. In contrast, Fig. 6.14 shows an abstract representation that allows users to focus on the relevant variables stored within the data set.

However, the visualisation of data sets is a secondary challenge, a follow-up to the primary challenge of data storage. Data sets included in a VGE often have a highly heterogeneous nature, varying in area of interest, extent, projection, structure and many other parameters.

The inclusion of modelling data and simulation results in the system is of vital importance for an effective analysis of processes. Often, different scenarios and their effect on the environment need to be displayed such that stakeholders or researchers are supported in their decision making process and can predict and understand consequences of changes to the environment. Examples include predictions about

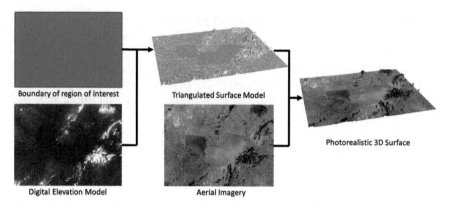

Fig. 6.7 Example of pre-processing steps necessary for the representation of 2D data sets in 3D space

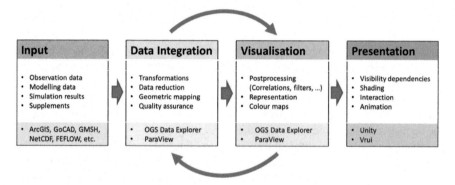

Fig. 6.8 Data integration workflow based on Helbig et al. (2017)

water management (e.g. short-term vs. long-term solutions) or waste water treatment (e.g. centralised vs. decentralised concepts). In addition to the challenges based on heterogeneity of the data, the inclusion of simulation requires the handling of (very) large data sets, the high number of parameters included, the transient structure of the data as well as the visualisation of uncertainty in simulated parameters.

To create a VGE that is not specifically tailored to one case study or region of interest requires flexible data structures and modular algorithms for storing, modifying and visualising data within the environment. This allows to use the same functionality for other case studies, even if these are focussed on different areas of application. The combination and parameterisation of these modules depend on the definition of a general workflows for data integration and visualisation (see Fig. 6.8) as well as flexible interfaces for connecting algorithms used within each of the steps within these workflows. The environmental information system as a whole is in turn embedded into a concept which provides a continuous data flow (Fig. 6.9) from the acquisition of data to the presentation of research results.

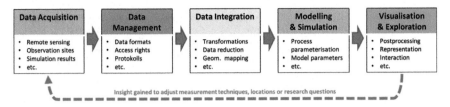

Fig. 6.9 Continuous modelling workflow concept based on Kolditz et al. (2012a)

6.3.3 Data Integration and Workflow

The import of all relevant data sets into a unified reference frame is handled by the OpenGeoSys Data Explorer (Rink et al. 2013). After the preprocessing stage, the integrated data sets are exported into the Unity engine (Goldstone 2011), which is used as the framework for the presented virtual geographic environment. The resulting applications can be used for presentations to audiences, such as stakeholders, researchers or the general public; but also for data exploration, discussions and verification of both input data and research results. Both programmes will be briefly introduced in the following.

OpenGeoSys Data Explorer

The Data Explorer is the graphical user interface to the platform-independent OpenGeoSys simulation software (Kolditz et al. 2012b). It is capable of importing and visualising measured data sets as well as simulation results. The framework already supports a large range of established file formats from the environmental sciences, incl. typical GIS formats (raster- or shapefiles), a variety of mesh generators for creating 3D objects, graphics formats and interfaces to a number of modelling and simulation frameworks. See Fig. 6.10 for an overview of currently supported file formats.

While OpenGeoSys is mostly used for the simulation of coupled Thermo-Hydro-Mechanical-Chemical (THMC) processes, the support of multi-purpose file formats such as VTK (Visualization Toolkit (Schroeder et al. 2006)) or NetCDF (Rew and Davis 1990) ensures that many simulation results from domains as diverse as continuum mechanics (e.g. OpenFOAM (Weller et al. 1998)), climate research (e.g. WRF (Michalakes et al. 2005)), lake and coastal research (e.g. GETM (Burchard and Bolding 2002)) or drainage network simulation (e.g. SWMM (Rossman 2014)) can be interpreted by the software. Measured or simulated parameter values (both static and transient) can be projected either on tessellated surfaces/volumes or geometric features.

Unity engine

Unity is a cross-platform game engine, developed for the creation of computer games for personal computers and mobile devices. The software is available for free and

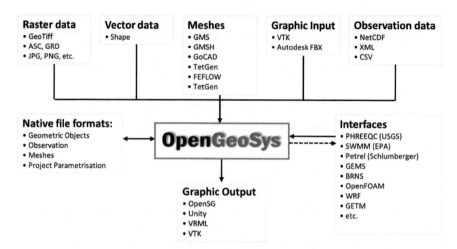

Fig. 6.10 Supported file formats for the OpenGeoSys framework. Due to the wide range of supported formats, interfacing with other simulation software is simple if an established format is supported. For instance, due to support of the NetCDF format, the integration of data from models such as the Weather Research & Forecast Model (WRF) used for climate simulations or the General Estuarine Transport Model (GETM) used for studies of lakes or coastal regions becomes straightforward

offers a wide range of features for displaying static and dynamic graphics in both 2D and 3D. Unity provides an interactive and extendable GUI that can be customised by users for creating certain types of applications. Dragging and dropping is supported for all types of Unity objects and it is possible to select, view and transform all data sets after import. Likewise, animation and interactive functionality can be instantly tested within the software without explicitly building an application. The software is extendable via the C# programming language, has a very large user base and a wide range of tutorials is available on the internet. Once functionality has been integrated, it is available within the Unity GUI and can subsequently be used to create new applications with that same set of methods.

Within the presented workflow, Unity is used in combination with the MiddleVR plug-in (MiddleVR Developers 2017). That way, a large range of technical challenges are automatically handled when building applications. This includes a wide range of rendering backends such as various operation systems for personal computers, web-applications, virtual reality (VR) environments or head-mounted displays; but also the required input devices for such environments, tracking of users in VR, or the support of various stereoscopic display modes.

Workflow

A typical workflow for integrating a new geoscientific data set into an existing virtual environment usually includes the following steps:

1. Import data set into Data Explorer framework
2. Create meaningful representations of the data set in 3D space
3. Detect inconsistencies between data sets (both visually and algorithmically)
4. If necessary, modify data set by adding missing information (such as elevation on 2D data, mapping of simulation results onto 3D objects, etc.) or correcting inconsistent or false information (e.g. topological incorrect surfaces, NaN values, etc.)
5. Present data in a meaningful fashion (e.g. choose adequate colours, apply textures, select glyphs or colour lookup tables for parameters of data points, etc.)
6. Add interaction methods to guide users to data set (camera paths, viewpoints, animations, etc.)

Steps 1–4 are usually performed using the OpenGeoSys Data Explorer. Once all data sets have been pre-processed, they are added to a scene generated within the Unity framework. Depending on structure and interpretation of each original data set, step 5 of our workflow may require algorithms implemented either in OpenGeoSys or in Unity. Step 6 is executed exclusively in Unity.

Being a game engine, the visualisation of environmental data sets is not a focus of the Unity framework. However, a lot of the functionality required for a VGE is already implemented in the software. This includes the the movement in 3D space, tracking of the user and the functionality to display and interact with complex objects. In addition, the framework is well-documented and easily extendable. We have developed a number of specific methods for importing and displaying geoscientific data sets, circumventing some of the limitations of the Unity engine. Geoscientific data sets are usually represented by objects that are larger then typical objects used in computer games and they rely more on a precise representation. Where computer games rely on automatically generated surfaces or roughly modelled objects that are displayed using computer graphics methods such as bump mapping or MIP mapping, geoscientific data has usually been exactly measured and does not contain repeatable features. We implemented algorithms for importing and visualising typical environmental data sets within Unity and added methods to deal with large data sets. Examples for such algorithms include the subdivision and use of very large textures (e.g. remote sensing imagery) or the automatic translation and scaling routines to convert the often very large coordinates from regional geographic projections into a small, scaled coordinate system typically used for scenes in graphics applications. Other application-specific issues include the use of modified shaders, e.g. for mapping pre-selected colour arrays on objects based on simulated parameters at varying time steps. For a more in-depth technical description the interested reader is referred to the article by Rink et al. (2017).

The Unity scene containing relevant data sets can be built as an executable file for a variety of target architectures, including virtual reality environments such as stereoscopic video walls or head-mounted displays (Oculus Rift, HTC Vive, etc.), but also for regular desktop computers. Users can navigate within the geographic environment with typical interaction devices. Besides keyboard and mouse this also includes typical VR interaction devices such as flysticks or gamepads. To make the

Fig. 6.11 The environmental information system for Chao Lake, displayed on the video wall in a Virtual Reality environment

framework easy to handle for inexperienced users, a number of features have been added for navigation and interaction:

- predefined viewpoints of relevant regions or data sets within the scene, such that users need not navigate the scene themselves and have pointers to regions of interest,
- the definition of camera paths along interesting features or gradients of natural or simulated parameters,
- picking objects to access supplemental content such as diagrams, photos, movies or webpages that give additional context to the currently viewed object or region,
- a simple animation interface to start, pause or step through time-variant content such as the time steps of simulation or remote sensing data, different scenarios, or predefined transformation of the data
- toggling of data sets or groups of data sets and adjusting their transparency to be able to focus on interesting aspects or compare data sets.

This functionality is accessible via a menu defined as on overlay in the 3D scene. The menu can be toggled via the interaction device to be visible/invisible and it can be moved across the view plane so it doesn't occlude interesting data. While it is possible to navigate freely in 3D space, experience has shown that most users prefer to access predefined points of view and interaction via a menu, similar to desktop applications. Figure 6.11 shows an example of a VGE based on case study of the Chao Lake in China, described in detail in the following section. The 3D visualisation is linked to time-series data measured from climate stations in the region and contains GIS data and simulation results as well as supplemental material such as pictures or diagrams. This facilitates a powerful combination of monitoring data and simulation results in one coherent, spatially arranged visualisation.

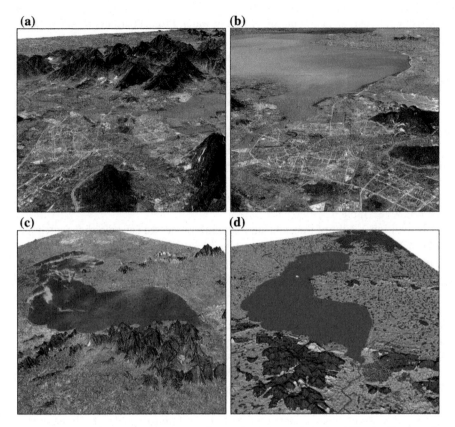

Fig. 6.12 Visualisation of the Chao Lake VGE. (**a**) Super-elevated terrain of Chaohu City from the north with aerial image texture. Buoys within the lake are marked in orange. (**b**) View of Chaohu City from the east with sewage network superimposed on the surface. Note the fine resolution of the texture within the city compared to the rougher texture in the background. (**c**) Super-elevated view of Chao Lake from the southwest. Mapped on the lake surface is a LANDSAT 8 image showing algal bloom in summer. (**d**) View of land use information in the Chao Lake region. Chaohu City is located in the lower left corner, road- and river network are mapped onto the terrain surface

6.3.4 Application Prototype

A Virtual Geographic Environment for the area around Chao Lake in China (see Fig. 6.12) has been set up as a prototype for the international project "Managing Water Resources for Urban Catchments" (Dohmann et al. 2016). The focus within this prototype is Chaohu, a city with a population of almost one million people, located on the eastern shore of Chao Lake within the Anhui Province in Eastern China. With a surface area of $720\,km^2$ Chao Lake is the fifth-largest freshwater lake in China. Due to eutrophication and silting, the lake has become one of the most polluted lakes in China in recent years (Chen et al. 2015a) which causes a number of problems for cities located at or near the lake, as the population uses the lake for fishing, irrigation and obtaining drinking water. Project partners from China and Germany are working together with the goal to analyse and improve the current situation at different stages of the fresh- and sewage water treatment workflow.

The purpose of the Virtual Geographic Environment for this region is to show the current situation in and around the lake, including risk factors such as algae or pesticide concentration. While the data collection is still ongoing, more than 30 data sets have been collected and processed into an immersive 3D scene to give users an impression of the region. Examples include digital elevation models (Tachikawa et al. 2011), a land cover map (Chen et al. 2015c), the bathymetry of Chao Lake, aerial imagery (Google Earth 2016), water bodies for the region of interest surrounding Chao Lake (OpenStreetMap contributors 2016) as well as detailed data on water and sewage networks within Chaohu City (Liao et al. 2016), etc. For a comprehensive overview, see Table 6.3.

In addition, results of transient simulation of processes have also been integrated into the system (see Table 6.4). A combined visualisation of monitoring and modelling data will support both stakeholders and researchers in their understanding and analysis of the complex coupled processes related to the pollution and management of water resources. In particular, a simulation of the drainage pipeline system for storm- and sewage water within the Chaohu City during rain events allows for an estimation of run-off times, the distribution of pollutants as well as the risk of flooding within the city (see Figs. 6.13 and 6.14). The implementation of the complex groundwater system model of the lake catchment shows the interaction with Chao Lake and the distribution of the algae toxin Microcystin (see Chap. 7). Finally, preliminary results of a lake simulation (Fig. 6.15) show temporal variation of temperature, evapotranspiration, etc. Similar to observation data, these simulation results can be explored in context with all the other data sets, with the additional benefit of user-controlled animation of the temporal variance of simulated parameters.

Our extensions of the Unity framework mentioned in the previous section are particularly useful for this case study. The overall region around Chao Lake integrated into the VGE has a size of over $5000\,km^2$. We have created a surface representation consisting of roughly 200,000 triangles with a maximum edge length of 250 m. The corresponding aerial imagery from Google (Google Earth 2016) and LANDSAT (Wulder et al. 2016) (see Fig. 6.12c) have a pixel size of roughly 12.5 m. The combination of those data sets results in a detailed visualisation of the surface data despite not incorporating advanced methods such as bump mapping (see Fig. 6.7 for a schematic of creating the textured terrain surface). However, the regions of interest for domain scientists are usually much smaller. For instance, for Chaohu City we built a refined surface representation with image data at much higher resolution. With a size of less than $100\,km^2$, this more detailed surface still consists of over 300,000 triangles with a maximum edge length of 10 m. The aerial imagery has a pixel size of less than 3 m. This fine resolution is necessary to maintain a useful representation of the area because users are expected to zoom in very close to the surface. Also, integrating very detailed data sets such as the drainage system for the city (with a sub-metre resolution) requires similarly refined corresponding data for a visual assessment of the relation of data sets to each other. Our extensions to the

Table 6.3 Observation-based data sets included in the environmental information system

Data type	Geospatial context	Data
Remote sensing data	Chao Lake catchment	ASTER Global Digital Elevation Model
		Google Earth aerial imagery
		LANDSAT8 imagery
		GLOBELAND30 land use data
	Chao Lake	Bathymetry
	Chaohu City	High resolution digital elevation model
		High resolution airial imagery
Vector data	Chao Lake	Catchment boundary
		OpenStreetMap data (waterbodies, roads, etc.)
		Lake boundary
		Buoy network
	Chaohu City	River network
		Roads
		Sewage network (Sewage pipelines, stormwater pipelines, outlets, etc.)
		Land development plan
	Shuagqiao He catchment	River and tributaries
		Subcatchments
Time-series data	Hefei City	Climate data 1952–2016 (Temperature, Precipitation, etc.)
	Chao Lake	Test buoy measurements
Supplemental data	Chao Lake	Photos
	Chaohu City	Traffic monitoring imagery
	Chaohu City water treatment plant	Photos
		Biomonitoring video
	Shuagqiao He catchment	Photos

Table 6.4 Model-based data sets included in the environmental information system

Software	Geospatial context	Data
OpenGeoSys	Chao Lake catchment	Structural model
		Groundwater head
		Groundwater flow direction
GETM	Chao Lake	Evaporation
		Heat flux
SWMM	Shuagqiao He catchment	Pollutant concentration
	Chaohu City	Sewage network

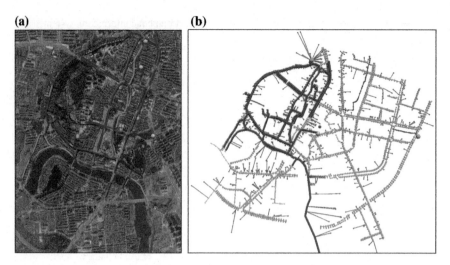

Fig. 6.13 Simulation of the Chaohu City sewage network using the Storm Water Management System (SWMM) (Rossman 2014). (**a**) SWMM Model within geographical context of Chaohu City. (**b**) Overlap of SWMM Model (in red) with the Chaohu City sewage network. All sewage data and simulation results have been provided by our project partners from Tongji University, Shanghai, China. SWMM output has then been imported into OpenGeoSys data structures for visualisation

Unity-framework allow to automatically choose to display the correct surface based on the position of the user. There will be automatic cross-fading between data sets when viewpoint within the city are selected from outside and vice versa.

When creating the VGE, data sets can be selected into groups, which is vital for a case study like this with many small data sets being part of a bigger picture. Both visibility and transparency can be adjusted simultaneously for all sets within each group, thus avoiding time-consuming selection or deselection of data sets. As an example, the case study contains 15 data sets contributing to the drainage pipeline system of Chaohu City (pipes, outlets, treatment plants, etc., see Fig. 6.16). Changing the status of each of these separately would not be viable during live presentations.

A number of dynamic data sets require the display of animations based on user input. For instance, time series of LANDSAT images of Chao Lake (Fig. 6.12c) visualises the annual algae growth. Likewise, results of transient simulations show the flow of groundwater, or the temporal distribution of pollutants within the sewage network, lake, or groundwater system. Especially the simulation of short term events, such as flooding, requires a very fine temporal resolution, often resulting in thousands of time steps. The example shown in Fig. 6.14 is taken from a simulation consisting of approximately 5000 time steps at an interval of five minutes.

While the virtual geographic environment of Chao Lake can be explored on a wide range of devices, as previously mentioned, it has been originally designed for use in a virtual reality environment using a 6 × 3 m video wall with additional projections at the sides and the floor where user position is tracked using infrared cameras (Bilke et al. 2014). Scientists from various research domains and stakeholders from the

6 WP-D: Environmental Information System

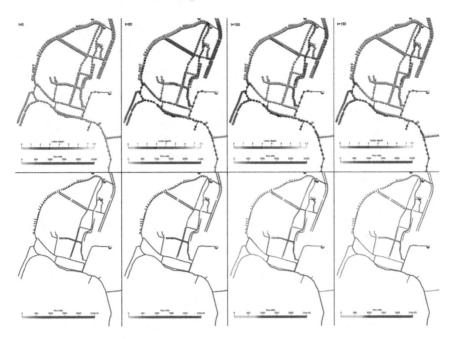

Fig. 6.14 Visualisation of a SWMM simulation of the Chaohu City sewage network during a rain event. Shown are the water depth (top line of the legend) with a range of 0–5 m and the flow rate (bottom line of the legend) of about 500 to 2000 m^3/s. With increased flow rate, water depth is critically increasing in certain parts of the network. Even after the flow rate has decreased again, it takes a long time for the water levels to reach normal levels again. The simulation results have been created by our project partners at Tongji University

German and Chinese Ministry of Science have participated in presentations and the response has been very positive. Complex contexts and interrelation of data sets are easy to understand even for users without a hydrogeological background and participants are often able to point out correlations or inconsistencies that would not have been visible within established software solutions, such as geographic information systems. While the VGE is still in a prototype stage, we hope to acquire more detailed data for the region in the future and extend the functionality offered by the system to import and display an even wider range of information and interaction.

6.3.5 Conclusions and Outlook

The presented work shows a generic framework for the development of Virtual Geographic Environments for Environmental Information Systems. This framework has been demonstrated for a case study of water resources management in the catchment of Chao Lake in the Anhui Province of China. All relevant types of data for water

Fig. 6.15 Embedded simulation of Chao Lake into 3D geographical scene. To enhance the existing topography of the region, the underlying DEM has been super-elevated by factor 5 and texturised using an airal image of the region. With a maximum depth of 5 m, Chao Lake is extremely shallow for a lake of its size. The depicted bathymetry has been super-elevated by factor 200 to give a sense of its 3D surface. Mapped on the lake surface is the result of preliminary simulation of evapotranspiration using the General Estuarine Transport Model (GETM) (Burchard and Bolding 2002)

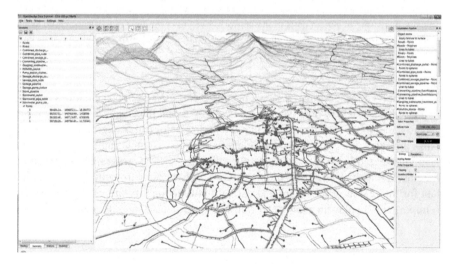

Fig. 6.16 Pre-processing of data sets. Depicted is the OpenGeoSys Data Explorer interface while integrating the drainage pipeline system in Chaohu City into the existing scene. The terrain surface mesh is semi-transparent visible in the background for reference as well as the road- (grey) and river network (blue). Fifteen data sets related to the drainage system have been loaded, mapped and assigned colours according to their function (e.g. violet data represents storm pipeline infrastructure, yellow signifies sewage infrastructure, etc.)

resources management that we are aware of have been integrated into the VGE with the help of both the OGS Data Explorer and the Unity game engine.

The presented VGE provides novelty in two directions: Firstly, static and dynamic data from both remote sensing and process modelling, can be integrated, explored and compared within the context of the VGE. Secondly, the framework allows for data integration at different spatial and temporal resolution, permitting to focus on specific regions of interest within a scene. The modular approach of the framework also allows to easily update existing data sets or add new data to an existing VGE.

Future activities are related to completing the data integration concept for all aquatic compartments (e.g. soil and groundwater systems) and to apply the framework to similar case studies within the Yangtze River Basin, such as the Poyang Lake or Tai Lake catchments. These case studies are also part of the "Research Centre for Environmental Information Science" (RCEIS - www.ufz.de/rceis).

References

Bilke L, Fischer T, and Helbig C et al. TESSIN VISLab - laboratory for scientific visualization. Environ. Earth Sci. **72**(10), 3881–3899 (2014). https://doi.org/10.1007/s12665-014-3785-5

Blöcher G, Cacace M, Reinsch T, and Watanabe N. Evaluation of three exploitation concepts for a deep geothermal system in the North German Basin. Comput. Geosci. **82**, 120–129 (2015)

Burchard H, and Bolding K. GETM – A General Estuarine Transport Model. Scientific Documentation. Technical Report EUR 20253 EN, European Commission (2002)

Chen Y, and Liu QQ. On the horizontal distribution of algal-bloom in Chaohu Lake and its formation process. Acta. Mech. Sin. **30**(5), 656–666 (2014)

Chen M, Lin H, Wen Y, He L, and Hu M. Sino-VirtualMoon: a 3D web platform using Chang'e-1 data for collaborative research. Planet. Space Sci. **65**, 130–136 (2012)

Chen M, Lin H, Wen Y, He L, and Hu M. Construction of a virtual lunar environment platform. Int. J. Digital Earth **6**(5), 469–482 (2013). https://doi.org/10.1080/17538947.2011.628415

Chen C, Börnick H, Cai Q, Dai X, Jähnig SC, Kong Y, Krebs P, Kuenzer C, Kunstmann H, Liu Y, Nixdorf E, Pang Z, Rode M, Schueth C, Song Y, Yue T, Zhou K, Zhang J, and Kolditz O. Challenges and opportunities of German-Chinese cooperation in water science and technology. Environ. Earth Sci. **73**(8), 4861–4871 (2015a). ISSN 1866-6299. https://doi.org/10.1007/s12665-015-4149-5

Chen M, Lin H, Kolditz O, and Chen C. Developing dynamic virtual geographic environments (VGEs) for geographic research. Environ. Earth Sci. **74**(10), 6975–6980 (2015b). ISSN 1866-6299. https://doi.org/10.1007/s12665-015-4761-4

Chen J, Chen J, and Liao A et al. Global land cover mapping at 30 m resolution: a POK-based operational approch. ISPRES J. Photogrammetry Remote Sens. **103**, 7–27 (2015c)

Dohmann M, Chen C, Grambow M, Kolditz O, Krebs P, Schmidt KR, Subklew G, Tiehm A, Wermter P, Dai XH, Liao ZL, Meng W, Song YH, Yin D, and Zheng BH. German contributions to the Major Water Program in China: "Innovation Cluster-Major Water". Environ. Earth Sci. **75**(8), 703 (2016). ISSN 1866-6299. https://doi.org/10.1007/s12665-016-5504-x

Fletcher K. Sentinel-3 – ESA's Global Land and Ocean Mission for GMES Operational Services. Technical Report ESA SP-1322/3, European Space Agency (2012)

Goldstone W. *Unity 3.x Game Development Essentials*, 2nd edn. (Packt Publishing, Birmingham, 2011)

Google Earth, Chao Lake, Anhui Province, China. Google Inc., Earthstar Geographics, CNES/Airbus DS, 2016. Accessed 26 Sept 2016

Helbig C, Bauer H-S, Rink K, Wulfmeyer V, Frank M, and Kolditz O. Concept and workflow for 3D visualization of atmospheric data in a virtual reality environment for analytical approaches. Environ. Earth Sci. **72**(10), 3767–3780 (2014). ISSN 1866-6299. https://doi.org/10.1007/s12665-014-3136-6

Helbig C, Dransch D, and Böttinger M et al. Challenges and strategies for the visual exploration of complex environmental data. Int. J. Digital Earth pp. 1–7 (2017). https://doi.org/10.1080/17538947.2017.1327618

Koch F, Bilke L, Helbig C, and Schlink U. Compact or cool? The impact of brownfield redevelopment on inner-city micro climate. Sustain. Cities Soc. **38**, 31–41 (2018). ISSN 2210-6707. https://doi.org/10.1016/j.scs.2017.11.021

Kolditz O, Rink K, and Shao H et al. International viewpoint and news: data and modelling platforms in environmental earth sciences. Environ. Earth Sci. **66**(4), 1279–1284 (2012a). https://doi.org/10.1007/s12665-012-1661-8

Kolditz O, Bauer S, and Bilke L et al. OpenGeoSys: an open source initiative for numerical simulation of thermo-hydro-mechanical/chemical (THM/C) processes in porous media. Environ. Earth Sci. **67**(2), 589–599 (2012b). https://doi.org/10.1007/s12665-012-1546-x

Li X-G, He H-Y, and Sun Q-F. The shallow groundwater pollutions assessment of west Liaohe plain (eastern). J. Chem. Pharm. Res. **5**(11), 290–295 (2013)

Liao Z, Zhi G, Zhou Y, Xu Z, and Rink K. To analyze the urban water pollution discharge system using the tracking and tracing approach. Environ. Earth Sci. **75**(14), 1080 (2016). ISSN 1866-6299. https://doi.org/10.1007/s12665-016-5881-1

Lin H, Chen M, and Lu G. Virtual geographic environment: a workspace for computer-aided geographic experiments. Ann. Assoc. Am. Geogr. **103**(3), 465–482 (2013a). https://doi.org/10.1080/00045608.2012.689234

Lin H, Chen M, Lu G, Zhu Q, Gong J, You X, Wen Y, Xu B, and Hu M. Virtual geographic environments (VGEs): a new generation of geographic analysis tool. Earth-Sci. Rev. **126**, 74–84 (2013b). ISSN 0012-8252. https://doi.org/10.1016/j.earscirev.2013.08.001

Lin H, Batty M, Jørgensen SE, Fu B, Konecny M, Voinov A, Torrens P, Lu G, Zhu A-X, Wilson JP, Gong J, Kolditz O, Bandrova T, and Chen M. Virtual environments begin to embrace process-based geographic analysis. Trans. GIS **19**(4), 493–498 (2015). ISSN 1467-9671. https://doi.org/10.1111/tgis.12167

Lü G. Geographic analysis-oriented virtual geographic environment: framework, structure and functions. Sci. China Earth Sci. **54**(5), 733–743 (2011). ISSN 1869-1897. https://doi.org/10.1007/s11430-011-4193-2

Michalakes J, Dudhia J, and Gill D. The weather research and forecast model: software architecture and performance, in *Proceedings of Eleventh ECMWF Workshop on the Use of High Performance Computing in Meteorology*. (World Scientific, Singapore, 2005), pp. 25–29. ISBN ISBN 978-9812563545

MiddleVR Developers. MiddleVR SDK – a generic immersive virtual reality plugin (2017). Accessed 15 Feb 2017

OpenStreetMap contributors. Planet dump retrieved from https://planet.osm.org (2016)

Rew R, and Davis G. NetCDF: an interface for scientific data access. IEEE Comput. Graphics Appl. **10**(4), 76–82 (1990)

Rink K, Bilke L, and Kolditz O. Setting up virtual geographic environments in unity, in *Proceedings of EuroVis Workshop on Visualization in Environmental Sciences*, pp. 1–5. EuroGraphics Digital Library (2017). ISBN 978-3-03868-040-6. https://doi.org/10.2312/envirvis.20171096

Rink K, Bilke L, and Kolditz O. Visualisation strategies for environmental modelling data. Environ. Earth Sci. **72**(10), 3857–3868 (2014). ISSN 1866-6299. https://doi.org/10.1007/s12665-013-2970-2

Rink K, Fischer T, Selle B, and Kolditz O. A data exploration framework for validation and setup of hydrological models. Environmental Earth Sciences **69**(2), 469–477 (2013). https://doi.org/10.1007/s12665-012-2030-3

Rossman L. SWMM-CAT User's Guide. Technical Report EPA 600-R-14-428, Environmental Protection Agency (2014)

Schroeder W, Martin K, and Lorensen B. *Visualization Toolkit: An Object-Oriented Approach to 3D Graphics*, 4th edn. (Kitware Inc., New York, 2006)

Tachikawa T, Kaku M, and Iwasaki A et al. ASTER Global Digital Elevation Model Version 2 - Summary of Validation Results. Technical report (NASA Jet Propulsion Laboratory, California Institute of Technology, 2011)

Tang DL, Kawamura H, Oh IS, and Baker J. Satellite evidence of harmful algal blooms and related oceanographic features in the Bohai Sea during autumn 1998. Adv. Space Res. **37**, 681–689 (2006)

Walther M, Bilke L, Delfs J-O, Graf T, Grundmann J, Kolditz O, and Liedl R. Assessing the saltwater remediation potential of a three-dimensional, heterogeneous, coastal aquifer system. Environ. Earth Sci. **72**(10), 3827–3837 (2014). ISSN 1866-6299. https://doi.org/10.1007/s12665-014-3253-2

Wang SF, Tang DL, and He FL et al. Occurrences of Harmful Algal Blooms (HABs) associated with ocean environments in the South China Sea. Hydrobiologia **596**, 79–93 (2008)

Weller HG, Tabor G, Jasak H, and Fureby C. A tensorial approach to computational continuum mechanics using object-oriented techniques. Comput. Phys. **12**(6), 620–631 (1998). ISSN 0894-1866. https://doi.org/10.1063/1.168744

Wulder MA, White JC, and Loveland T et al. The global landsat archive: status, consolidation, and direction. Remote Sens. Environ. **185**, 271–283 (2016)

Chapter 7
WP-E: Groundwater Systems

Martin Pohl, Christian Engelmann and Marc Walther

7.1 Introduction and Goals

The work package E was introduced as an addition to the "Urban Catchments"-project, in late 2016, more than one and a half year after its start. Its objective was to create a groundwater model to complete the hydraulic system which so far was limited to the urban and suburban regions around Chao Lake and the lake itself. The vast network of rivers, streams and ditches in combination with the domination of agricultural land use in the area around Chao Lake results in an inflow of nutrients and possibly other pollutants by surface water. But it is assumed that a considerable amount of pollutants and nutrients find their way into Chao Lake via groundwater flow, after they presumably leaked from defect sewage pipes or seeped into the groundwater directly from their source at the surface.

Another interesting research topic that was picked up in work package E can be attributed to a scientific paper by Yang et al. (2016c). In this paper they presented the results of their research about the occurrence of Microcystins in the groundwater

---

M. Pohl (✉) · C. Engelmann
Department of Hydrosciences, Institute for Groundwater Management,
Technische Universität Dresden, Bergstraße 66, 01069 Dresden, Germany
e-mail: Martin.Pohl@mailbox.tu-dresden.de

C. Engelmann
e-mail: Christian.Engelmann@tu-dresden.de

M. Walther
Department of Environmental Informatics,
Helmholtz Centre of Environmental Research–UFZ,
Permoserstr. 15, 04318 Leipzig, Germany

M. Walther
Department of Hydrosciences, Institute for Groundwater Management,
Professorship of Contaminant Hydrology, Technische Universität Dresden,
Bergstraße 66, 01069 Dresden, Germany
e-mail: marc.walther@ufz.de

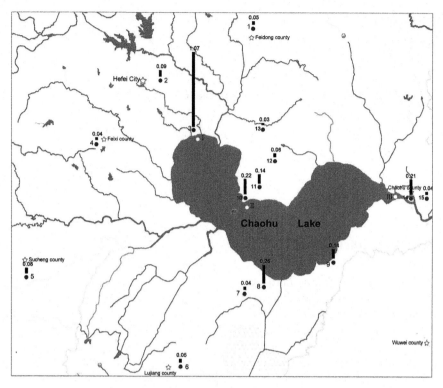

Fig. 7.1 Sampling locations in Chao Lake watershed and their Microcystin concentrations in September 2013 (Yang et al. 2016c)

around Chao Lake. Microcystins are a group of hepatotoxins which can be produced by a number of cyanobacteria and are a recurring threat for the population around Chao Lake because of cyanobacterial blooms. According to Yang et al. (2016c) it apparently has leaked from the lake into the surrounding groundwater as can be seen in Fig. 7.1. While the found concentrations of Microcystins were mostly but not entirely below the provisional guideline value for Microcystin-LR of 1 µg/L issued by WHO (1998), the risk posed by lower concentrations is still a topic of research. The International Agency for Research on Cancer (IARC) has classified Microcystin-LR, a structural variant of the Microcystins, as possibly carcinogenic and points out that there is strong evidence that it is a tumor promotor (IARC Working Group 2010).

A highlight during the course of the work on work package E was the meeting with the team lead by Prof. Weiping Hu, our partner from the Nanjing Institute of Geography and Limnology, Chinese Academy of Science (NIGLAS). The meeting took place in the city of Nanjing in May 2017. There we presented the state of our work on the groundwater model at that time and discussed further cooperation. As a result of this cooperation we were able to access a significant amount of data that helped to increase the quality of the model and allowed to progress from a steady state model to a transient model. In addition the need for further monitoring in the region regarding the groundwater situation was noted.

7.2 Data and Tools

The creation of the groundwater model for the region around Chao Lake, as described in this Chapter, required the utilization of a number of tools and the incorporation of a large amount of data. At the centre of the model development process is the open source simulation program OpenGeoSys (OGS). OGS offers a method for the numerical simulation of thermo-hydro-mechanical-chemical processes in porous and fractured media (Kolditz et al. 2012a). Therefore, the preprocessing of the available data not only includes a general information extraction but also the conversion to an input format that is compatible with OGS. The data used contains large scale raster files that cover the whole model area but also data from measuring stations and from samplings which consequently contain spatial limited information. Additionally, values for process variables were researched from literature, such as for Microcystin mass transport. For its implementation a value for degradation and a sorption isotherm are necessary.

7.2.1 Data

Data on the topography were mainly obtained from two sources. On the one hand, this is the Digital Elevation Model (DEM) SRTM, which stands for Shuttle Radar Topography Mission (Farr et al. 2007). The DEM was derived from radar interferometer data recorded during Space Shuttle Missions and has a resolution of three arc seconds or approximately 90 meters. Earlier precipitation data contains large gaps and some of the water level time series don't reach further back. Furthermore there is no data for some of the rivers sections, especially in the upstream direction.

7.2.2 Tools

For the purposes of preprocessing a number of different tools were applied. A tool that was frequently used was the geographic information system Quantum-GIS (QGIS). Similarly the text editor Notepad++ and the spreadsheet software Excel were important tools at multiple steps. Other used software were the OGS Data Explorer, GINA_OGS and the the finite element mesh generator GMSH.

7.3 Methods

7.3.1 Model Geometry (Outer Boundary; Lake and Rivers)

The outer boundary of the groundwater model was deduced from the digital elevation model data to represent the surface catchment area of Chao Lake. The surface catchment area was chosen as a surrogate for the groundwater catchment, since the hydrogeological data did not allow to identify a different boundary. There are a number of scientific papers available that include figures of the surface catchment area, although the depicted boundaries vary in parts from one scientific paper to another. While those figures can't be used directly, they can serve as guidance and validation for the outer model boundary that was calculated.

The calculation was done in the geographic information system Quantum-GIS (QGIS)[1] by using the GRASS-GIS tool "r.watershed". The underlying algorithm can determine basins based on the digital elevation model (DEM) raster data and QGIS outputs the result in form of polygons which delineate the watersheds. Minor adjustments were necessary since bridges and other structures distort the elevation data and are mistaken as watersheds by the algorithm. This happened prevalently for areas with low slope where the flow direction is easily changed by little flaws in the elevation data.

The resulting catchment area, which can be seen in Fig. 7.2 as a red line, is characterized by the Yangtze River to the east, a mountainous area in the southwest which goes up to around 1300 m and a relatively flat area around Chao Lake. Some of the bigger rivers that flow into the lake and the only river flowing out of it have been included into the model. It was assumed that these surface water bodies likely influence the groundwater flow. The region is generally abundant in streams, ditches, reservoirs and ponds, but for the purpose of limiting the complexity of the model only a selection of them were included.

7.3.2 Mesh (Generating; Mesh Properties (Material Class))

The basic geometric components and the topography data allows for the creation of the finite element mesh which is the base frame for the model. The creation of a two-dimensional mesh was done by employing the mesh generator GMSH.[2] At that it will integrate the geometric points of the watershed and river polylines as nodes into the mesh. It is necessary to account for the fact that the level of detail of the mesh influences the quality and detail of the model results but also the time that is required to solve the model simulation. Especially mass transport processes require a high spatial resolution to ensure a stable solution. A suitable tradeoff has

---

[1] https://www.qgis.org/de/site/.
[2] http://gmsh.info/.

7 WP-E: Groundwater Systems

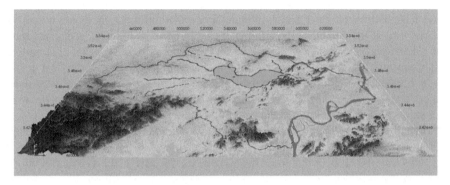

Fig. 7.2 Digital elevation model with Chao Lake catchment boundary (red line) and a number of surface water bodies (blue lines), visualized in paraview

Fig. 7.3 Finite element mesh based on a simplified geometry (left) and a cross-section of a three dimensional mesh (right). Colored according to the finite elements corresponding material group

to be made and therefore the polylines were simplified beforehand, so that they are composed of few hundred points instead of the around original ten thousand. The difference can easily been seen by comparing Fig. 7.2 where the geometry polylines are very detailed, with Fig. 7.3 where the geometry polylines are composed of long and straight components. At the same time the sizes of the finite elements closer to Chao Lake were set to allow for more accurate simulation results by increasing the number of nodes that are created during the mesh generation process.

The two-dimensional mesh was further processed in the OGS Data Explorer by adding the elevation data from the DEM to the mesh nodes. Since the currently available hydrogeological data didn't contain detailed information about the spatial variability of the depths of the aquifer the simplifying assumption was made that the depth was constant over the whole model domain. Based on that assumption subsequently layers were added by utilizing the Data Explorer until the final depth was reached. Thereby the two-dimensional mesh changed to a three-dimensional mesh and the finite elements changed from triangles to prisms. Similarly to the two-dimensional mesh, the sizes of the new finite elements were set for better accuracy closer to the Lake. This was achieved by having thinner upper layers and thicker bottom layers.

The material class of the upper layer elements was assigned according to information that was derived from a groundwater resources distribution map of Anhui province which was found in the digital library of the National Geological Archive of China.[3] Sedimentary materials with higher hydraulic conductivity occur mainly alongside the river courses, as can be seen in Fig. 7.3 where those materials are colored in green. They are likely of fluvial origin. Less conductive loose rock is colored in yellow and karst rock in blue. The red colored areas represents fractured rock materials and corresponds with higher elevations. A different material class was homogeneously assigned to the finite elements of the lower layers, based on the available hydrogeological data.

7.3.3 Boundary Conditions and Source Terms (Time Series; Water Levels; Recharge)

With the finite element mesh in place boundary conditions and source terms can be assigned to the nodes. For this purpose a large amount of time series data was preprocessed in a spreadsheet software and subsequently assigned to the corresponding mesh node. The data consisted of water level time series, for example for Chao Lake and Yangtze River, and precipitation data for a number of precipitation recording stations throughout Anhui Province. It was partly provided by our Chinese partners from NIGLAS and partly obtained from the online portal of the Bureau of Hydrology of the Province of Anhui.[4] Taking into account the consistency of the time series data, a period of around ten years was chosen for the simulation, starting with the first of July 2007. Five-day-averages were calculated to reduce the amount of the time series data that had to be integrated into the model as boundary condition or as source term.

River water levels between two measuring sites or between a measuring site and Chao Lake were extrapolated for each five-day-average value. When no information about the water level was available, which concerned some upstream segments of rivers, the surface elevation was used instead. It was ensured that this didn't lead to unrealistic behavior like a river flowing in two different directions at the same time.

---

[3]http://en.ngac.org.cn/Document/Map.aspx?MapId=EC7E1A7A7C331954E0430100007F182E.
[4]http://yc.wswj.net/ahyc/.

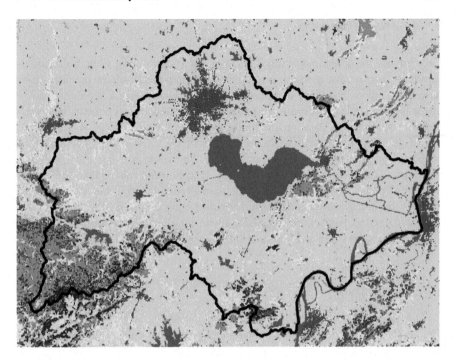

Fig. 7.4 Land cover in the year 2015 according to the ESA CCI data, visualized in QGIS. © ESA climate change initiative - land cover project 2017

The model area was divided into nine subareas along county and district borders, to accommodate for spatial variety in precipitation. Since several precipitation recording stations were located in and close to each subarea the arithmetic average was calculated to obtain a single precipitation time series for each subarea. For the time periods when no data was available for a subarea, data from the nearest active recording stations was used instead. The actual groundwater recharge by precipitation is applied as source term to the surface mesh nodes and is composed of the precipitation, the size of the area associated with a node and a recharge rate which describes how much of the precipitation actually infiltrates into the aquifer. The recharge rates were chosen based on available literature values. Recharge rates vary depending on the land cover, therefore land cover data from the ESA CCI Land Cover Project,[5] as it can be seen in Fig. 7.4, was used to assign the respective recharge rates to the nodes. The biggest impact can be expected from the recharge factor for cropland since cropland amounts to around three quarters of the land cover in the model area, with the rest mainly being tree cover, water bodies and urban areas.

---

[5] http://maps.elie.ucl.ac.be/CCI/viewer/index.php.

7.3.4 Mass Transport

To reproduce the spreading of the algae toxin Microcystin from Chao Lake into the surrounding groundwater three mass transport components where included in the model in form of boundary conditions. One component represents a conservative tracer and is intended to serve as a reference to the other two components. Those two components represent sorption and biodegradation of Microcystin-LR respectively, one of the most researched structure variants of the toxin group. The separation of those two properties shall help to identify their respective significance for the transport process. The literature values for the sorption constant and the biodegradation rate depend on multiple factors. The biodegradation rate for example depends among other things on what bacterium strains are involved. Since Microcystin was measured in great distances to Chao Lake literature values were chosen that promote the spreading thus the choice fell on a small sorption constant and a low decay rate.

7.3.5 Simulations

The complexity of the different model simulations increased in the course of the project and with the amount of available data. Initially a steady-state two dimensional model was created with Chao Lake, Yangtze River and a number of smaller rivers as surface water level boundary conditions. At this point the respective water levels were deduced from the elevation data of the SRTM Digital Elevation Model and the groundwater recharge by precipitation was derived from yearly average precipitation values available for Chaohu City. Next up the two-dimensional model was extended by one dimension to be able to identify differences in flow depending on the depth. After receiving time series data for precipitation and surface water levels, the steady-state model was further developed into a transient model. The simulation encompasses a ten year period beginning with the first of July 2007. Subsequently mass transport components where added, further increasing the model complexity. The time needed to complete a simulation run increases with the complexity of the model. While the steady-state models took seconds to a few minutes for the completion of the simulation, the three dimensional ten year transient model with mass transport took days to a few weeks, highly depending on the number of included mass transport components.

7.3.6 Visualization and Analysis

For the purpose of visualization and further analysis the program ParaView was used. The results of the simulation in OpenGeoSys were saved as Visualization Toolkit Format (VTK) file, which is compatible with ParaView.

7.4 Preliminary Results

7.4.1 Two-Dimensional Steady-State Model

During the early stages of the work on the groundwater model a two-dimensional steady-state model was created. At this point the time series data for water levels and precipitation weren't available yet, plus some of the hydrogeological data. Therefore surrogates were used in place of the missing information. The surface elevation data from the SRTM was used to represent the surface water level boundary conditions at the respective locations. The precipitation and the groundwater recharge by precipitation respectively were deduced from available values for the average yearly precipitation. While taking this limitations into consideration the model allowed for a first insight.

The groundwater flow in the model mainly and expectably follows the natural slope in direction of Chao Lake, as can be seen in Fig. 7.5. However the flow is affected by the river boundary conditions which results in some of the visible velocity vectors in Fig. 7.5 being deviated towards the rivers. In the southeast the influence of the Yangtze river is especially apparent as the groundwater flow is divided halfway to the lake. The influence of the Yuxi river, which flows from Chao Lake into the Yangtze, can be seen in the northeast where the groundwater flow direction leads into this river.

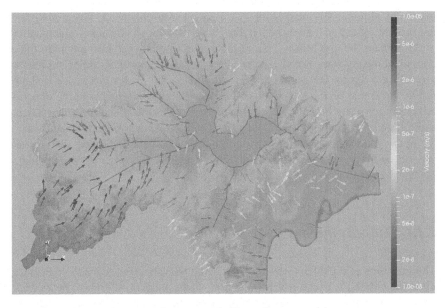

Fig. 7.5 Groundwater flow according to an early 2D steady-state model

7.4.2 Three-Dimensional Steady-State Model

Further into the project the two-dimensional model was expanded to a three-dimensional model which made it possible to factor in the vertical movement of the groundwater and later on also the vertical movement of the mass transport components. This includes movement underneath the lake and the river boundary conditions. At that time additional hydrogeological data for the upper layers was obtained and the expansion in the third dimension allowed to create layers with different hydrogeological properties.

7.4.3 Transient Model

With the receipt of the time series data for the surface water levels and precipitation the model and the simulation results became more dynamic. The simulated time period comprised approximately 10 years, starting with the first of July 2007. With the integration of variable water levels and precipitation the interaction between Chao Lake and the adjacent subsurface also became varying. According to the data the differences in precipitation and surface water levels throughout the year are strong. Shortly before the rainy season the water level of Chao Lake usually reaches its minimum for the year, which is likely a result of active management to prepare for the rain. At its peak the water level difference can be several meters compared to the minimum, although for most years it's a little bit over one meter. The groundwater levels in the model on the other hand don't react as strong to the heavy precipitation.

As a result the groundwater isn't only exfiltrating into the lake anymore like it does in the steady-state model, but the lake water also temporarily infiltrates into the groundwater during the course of year. An illustration of this difference is shown in Fig. 7.6 where on the left side the velocity vectors point mostly in the direction of the surface water boundary condition at the model surface and therefore indicate exfiltration of groundwater into Chao Lake, while the vectors on the right side of the figure point away from the surface into the subsurface, indicating that the lake water is infiltrating in the underground. Similar activity was observed at the Yangtze boundary condition.

7.4.4 Mass Transport

The spreading of the three mass transport components varies according to the magnitude and direction of the groundwater flow. There are also differences between the components in regards to the extent of the spreading. These differences are apparent when looking at Fig. 7.7, which shows how far the three components have spread at day 1120 of the simulation by means of a cross section through Chao Lake. At

7 WP-E: Groundwater Systems   241

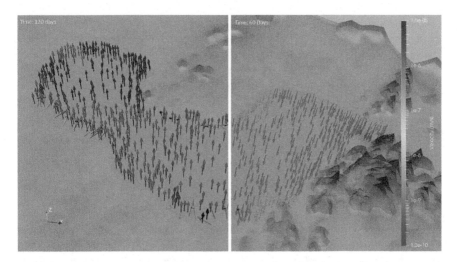

Fig. 7.6 Groundwater exfiltrating into Chao Lake at 120 days (left). Water from Chao Lake infiltrating into the groundwater at 60 days (right)

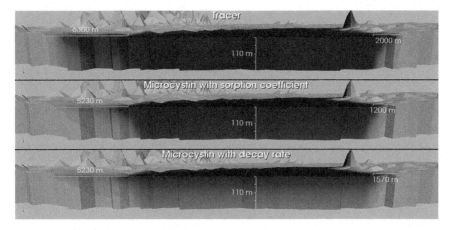

Fig. 7.7 Comparison of the spreading (red colour) of the three mass transport components by means of a cross section at 1120 days

this point the water level of the lake is particularly high. The Tracer component has spread further compared to the two Microcystin components. In the layers closer to the surface the Microcystin Decay component has spread further than the Sorption component since it isn't affected by sorption. In return the sorption component is more persistent. When the water level of Chao Lake lowers and the resulting change in the groundwater flow direction reduces or stops the mass transport components from further leaking from the lake, the remaining Microcystin Decay component is quickly reduced.

Under consideration of the available input data the results of the model simulation are in agreement with Yang et al. (2016c) who reported that the groundwater around Chao Lake was contaminated with Microcystin and that it is spread by lake water infiltrating into the groundwater. However the model has to be further improved to reproduce the reported spatial distribution of Microcystin.

7.5 Outlook

As was already mentioned in the introduction the groundwater systems work package is an addition to the "Urban Catchments"-project that was added more than one and a half year after the project start. Since then a highly informative three-dimensional transient groundwater model was created, based on the available data. The evaluation of the simulation results has just begun and there is potential for more precise simulations if more detailed hydrogeological data can be obtained. There is also potential for research into the transport of the algae toxin Microcystin in the subsurface and in groundwater.

Acknowledgements We acknowledge the ESA CCI Land Cover project for providing land cover data for the Chao Lake catchment (ESA Climate Change Initiative - Land Cover project 2017): http://maps.elie.ucl.ac.be/CCI/viewer/download/ESACCI-LC-Ph2-PUGv2_2.0.pdf.

References

Farr TG, Rosen PA, Caro E, Crippen R, Duren R, Hensley S, Kobrick M, Paller M, Rodriguez E, and Roth L et al. The shuttle radar topography mission. Rev. Geophys. **45**(2) (2007)

IARC Working Group on the Evaluation of Carcinogenic Risk to Humans, editor. *IARC Monographs on the Evaluation of Carcinogenic Risks to Humans: Ingested Nitrate and Nitrite, and Cyanobacterial Peptide Toxins.*, vol. 94. International Agency for Research on Cancer, Lyon (FR), 2010

Kolditz O, Bauer S, Bilke L, öttcher NB, Delfs J-O, Fischer T, Görke UJ, Kalbacher T, Kosakowski G, and McDermott CI et al. OpenGeoSys: an open-source initiative for numerical simulation of thermo-hydro-mechanical/chemical (THM/C) processes in porous media. Environ. Earth Sci. **67**(2), 589–599 (2012a)

W. H. O. WHO. Guidelines for drinking-water quality, 2nd edn. vol. 2. Health Criteria and Other Supporting Information. Addendum (1998)

Yang Z, Kong F, and Zhang M. Groundwater contamination by microcystin from toxic cyanobacteria blooms in Lake Chaohu, China. Environ. Monit. Assess. **188**(5), 280 (2016c). ISSN 0167-6369, https://doi.org/10.1007/s10661-016-5289-0

Printed by Printforce, the Netherlands